LIE ALGEBRAS in PARTICLE PHYSICS

From Isospin to Unified Theories

FRONTIERS IN PHYSICS: A Lecture Note and Reprint Series

David Pines, Editor

Volumes of the Series published from 1961 to 1973 are not officially numbered. The parenthetical numbers shown are designed to aid librarians and bibliographers to check the completeness of their holdings.

(1)	N. Bloembergen	Nuclear Magnetic Relaxation: A Reprint Volume, 1961
(2)	G. F. Chew	*S*-Matrix Theory of Strong Interactions: A Lecture Note and Reprint Volume, 1961
(3)	R. P. Feynman	Quantum Electrodynamics: A Lecture Note and Reprint Volume, 1961 (6th printing, 1980)
(4)	R. P. Feynman	The Theory of Fundamental Processes: A Lecture Note Volume, 1961 (6th printing, 1980)
(5)	L. Van Hove, N. M. Hugenholtz, and L. P. Howland	Problems in Quantum Theory of Many-Particle Systems: A Lecture Note and Reprint Volume, 1961
(6)	D. Pines	The Many-Body Problem: A Lecture Note and Reprint Volume, 1961 (5th printing, 1979)
(7)	H. Frauenfelder	The Mössbauer Effect: A Review—with a Collection of Reprints, 1962
(8)	L. P. Kadanoff and G. Baym	Quantum Statistical Mechanics: Green's Function Methods in Equilibrium and Nonequilibrium Problems, 1962 (5th printing, 1981)
(9)	G. E. Pake	Paramagnetic Resonance: An Introductory Monograph, 1962 [cr. (42)—2nd edition]
(10)	P. W. Anderson	Concepts in Solids: Lectures on the Theory of Solids, 1963 (4th printing, 1978)
(11)	S. C. Frautschi	Regge Poles and *S*-Matrix Theory, 1963
(12)	R. Hofstadter	Electron Scattering and Nuclear and Nucleon Structure: A Collection of Reprints with an Introduction, 1963
(13)	A. M. Lane	Nuclear Theory: Pairing Force Correlations to Collective Motion, 1964
(14)	R. Omnès and M. Froissart	Mandelstam Theory and Regge Poles: An Introduction for Experimentalists, 1963
(15)	E. J. Squires	Complex Angular Momenta and Particle Physics: A Lecture Note and Reprint Volume, 1963
(16)	H. L. Frisch and J. L. Lebowitz	The Equilibrium Theory of Classical Fluids: A Lecture Note and Reprint Volume, 1964
(17)	M. Gell-Mann and Y. Ne'eman	The Eightfold Way: (A Review—with a Collection of Reprints), 1964
(18)	M. Jacob and G. F. Chew	Strong-Interaction Physics: A Lecture Note Volume, 1964
(19)	P. Nozières	Theory of Interacting Fermi Systems, 1964

FRONTIERS IN PHYSICS: A Lecture Note and Reprint Series

David Pines, Editor *(continued)*

(20)	J. R. Schrieffer	Theory of Superconductivity, 1964 (2nd printing, 1971)
(21)	N. Bloembergen	Nonlinear Optics: A Lecture Note and Reprint Volume, 1965 (3rd printing, with addenda and corrections, 1977)
(22)	R. Brout	Phase Transitions, 1965
(23)	I. M. Khalatnikov	An Introduction to the Theory of Superfluidity, 1965
(24)	P. G. de Gennes	Superconductivity of Metals and Alloys, 1966
(25)	W. A. Harrison	Pseudopotentials in the Theory of Metals, 1966 (2nd printing, 1971)
(26)	V. Barger and D. Cline	Phenomenological Theories of High Energy Scattering: An Experimental Evaluation, 1967
(27)	P. Choquard	The Anharmonic Crystal, 1967
(28)	T. Loucks	Augmented Plane Wave Method: A Guide to Performing Electronic Structure Calculations—A Lecture Note and Reprint Volume, 1967
(29)	T. Ne'eman	Algebraic Theory of Particle Physics: Hadron Dynamics in Terms of Unitary Spin Currents, 1967
(30)	S. L. Adler and R. F. Dashen	Current Algebras and Applications to Particle Physics, 1968
(31)	A. B. Migdal	Nuclear Theory: The Quasiparticle Method, 1968
(32)	J. J. J. Kokkedee	The Quark Model, 1969
(33)	A. B. Migdal and V. Krainov	Approximation Methods in Quantum Mechanics, 1969
(34)	R. Z. Sagdeev and A. A. Galeev	Nonlinear Plasma Theory, 1969
(35)	J. Schwinger	Quantum Kinematics and Dynamics, 1970
(36)	R. P. Feynman	Statistical Mechanics: A Set of Lectures, 1972 (6th printing, 1981)
(37)	R. P. Feynman	Photo-Hadron Interactions, 1972
(38)	E. R. Caianiello	Combinatorics and Renormalization in Quantum Field Theory, 1973
(39)	G. B. Field, H. Arp, and J. N. Bahcall	The Redshift Controversy, 1973 (2nd printing, 1976)
(40)	D. Horn and F. Zachariasen	Hadron Physics at Very High Energies, 1973
(41)	S. Ichimaru	Basic Principles of Plasma Physics: A Statistical Approach, 1973 (2nd printing, with revisions, 1980)
(42)	G. E. Pake and T. L. Estle	The Physical Principles of Electron Paramagnetic Resonance, 2nd Edition, completely revised, enlarged, and reset, 1973 [cf. (9)—1st edition]

FRONTIERS IN PHYSICS: A Lecture Note and Reprint Series

David Pines, Editor *(continued)*

Volumes published from 1974 onward are being numbered as an integral part of the bibliography:

Number		
43	R. C. Davidson	Theory of Nonneutral Plasmas, 1974
44	S. Doniach and E. H. Sondheimer	Green's Functions for Solid State Physicists, 1974 (2nd printing, 1978)
45	P. H. Frampton	Dual Resonance Models, 1974
46	S. K. Ma	Modern Theory of Critical Phenomena, 1976 (4th printing, 1981)
47	D. Forster	Hydrodynamic Fluctuations, Broken Symmetry, and Correlation Functions, 1975 (2nd printing, 1980)
48	A. B. Migdal	Qualitative Methods in Quantum Theory, 1977
49	S. W. Lovesey	Condensed Matter Physics: Dynamic Correlations, 1980
50	L. D. Faddeev and A. A. Slavnov	Gauge Fields: Introduction to Quantum Theory, 1980
51	P. Ramond	Field Theory: A Modern Primer, 1981
52	R. A. Broglia and A. Winther	Heavy Ion Reactions: Lecture Notes Vol. I: Elastic and Inelastic Reactions, 1981
53	R. A. Broglia and A. Winther	Heavy Ion Reactions: Lecture Notes Vol. II, *in preparation*
54	Howard Georgi	Lie Algebras in Particle Physics: From Isospin to Unified Theories, 1982

Other Numbers in preparation

LIE ALGEBRAS in PARTICLE PHYSICS

From Isospin to Unified Theories

Howard Georgi
Lyman Laboratory of Physics
Harvard University
Cambridge, Massachusetts

Introduction by
Sheldon L. Glashow
Harvard University

1982
THE BENJAMIN/CUMMINGS PUBLISHING COMPANY, INC.
ADVANCED BOOK PROGRAM
Reading, Massachusetts

London · Amsterdam · Don Mills, Ontario · Sydney · Tokyo

CODEN: FRPHA

Library of Congress Cataloging in Publication Data

Georgi, Howard.
Lie algebras in particle physics.
(Frontiers in physics; v. 54)
Bibliography: p.
Includes index.
1. Lie algebras. 2. Particles (Nuclear physics)
3. S-matrix theory. I. Title. II. Series.
QC793.3.M36G46 539.7'0151255 81-18106
ISBN 0-8053-3153-0 AACR2

Reproduced by Benjamin/Cummings Publishing Company, Inc., Advanced Book Program, Reading Massachusetts, from camera-ready copy prepared by the author.

Manufactured in the United States of America

ABCDEFGHIJ-HA-8987654321

TO MY PARENTS

CONTENTS

EDITOR'S FOREWORD

The problem of communicating in a coherent fashion recent developments in the most exciting and active fields of physics seems particularly pressing today. The enormous growth in the number of physicists has tended to make the familiar channels of communication considerably less effective. It has become increasingly difficult for experts in a given field to keep up with the current literature; the novice can only be confused. What is needed is both a consistent account of a field and the presentation of a definite "point of view" concerning it. Formal monographs cannot meet such a need in a rapidly developing field, and, perhaps more important, the review article seems to have fallen into disfavor. Indeed, it would seem that the people most actively engaged in developing a given field are the people least likely to write at length about it.

FRONTIERS IN PHYSICS has been conceived in an effort to improve the situation in several ways. Leading physicists today frequently give a series of lectures, a graduate seminar, or a graduate course in their special fields of interest. Such lectures serve to summarize the present status of a rapidly developing field and may well constitute the only coherent account available at the time. Often, notes on lectures exist (prepared by the lecturer himself, by graduate students, or by postdoctoral fellows) and are distributed in mimeographed form on a limited basis. One of the principal purposes of the FRONTIERS IN PHYSICS Series is to make such notes available to a wider audience of physicists.

It should be emphasized that lecture notes are necessarily rough and informal, both in style and content;

and those in the series will prove no exception. This is as it should be. The point of the series is to offer new, rapid, more informal, and, it is hoped, more effective ways for physicists to teach one another. The point is lost if only elegant notes qualify.

The publication of collections of reprints of recent articles in very active fields of physics will improve communication. Such collections are themselves useful to people working in the field. The value of the reprints will, however, be enhanced if the collection is accompanied by an introduction of moderate length which will serve to tie the collection together and, necessarily, constitute a brief survey of the present status of the field. Again, it is appropriate that such an introduction be informal, in keeping with the active character of the field.

The informal monograph, representing an intermediate step between lecture notes and formal monographs, offers an author the opportunity to present his views of a field which has developed to the point where a summation might prove extraordinarily fruitful but a formal monograph might not be feasible or desirable.

Contemporary classics constitute a particularly valuable approach to the teaching and learning of physics today. Here one thinks of fields that lie at the heart of much of present-day research, but whose essentials are by now well understood, such as quantum electrodynamics or magnetic resonance. In such fields some of the best pedagogical material is not readily available, either because it consists of papers long out of print or lectures that have never been published.

The above words, written in 1961, seem equally applicable today. It was in that year that Murray Gell-Mann and Yuval Ne'eman found the symmetry group {SU(3), now called flavor SU(3)} which made possible the organization of existing elementary particles into a meaningful pattern (the eight-fold way), and led to the discovery of the Ω^- particle. As Sheldon Glashow notes in his eloquent Introduction to this volume, from that time on group theory, and especially the theory of Lie groups, has been an indispensable tool of the particle theorist. Howard Georgi, as the co-inventor (with Glashow) of the group SU(5), has superb credentials for writing the present volume, and I share his hope that it will make the theory of Lie groups accessible to graduate students while offering a perspective on the way in which knowledge of such groups can provide insight into the development of unified theories of strong, weak, and electromagnetic interactions.

Howard Georgi notes in his Preface that it was a FRONTIERS IN PHYSICS volume (Richard Feynman, *The Theory of Fundamental Processes*) which kindled his own interest in theoretical physics. It is therefore a special pleasure to welcome him as a "second-generation" contributor to this Series, and to express the hope that this volume, in turn, might attract other talented young scientists to theoretical physics, and, in time, to the ranks of authors in this Series.

David Pines

INTRODUCTION

My graduate education at Harvard began in 1954 with John Van Vleck's famous course in group theory for physicists. Relics of the course -- multiplication tables for small finite groups -- still adorn the lecture halls of Jefferson Laboratory. The course has changed in the succeeding quarter century, passing through many hands (mine included) to arrive at its present form under the direction of Professor Howard Georgi. I am proud to have convinced Howard to publish his valuable lecture notes. The evolution of the course reflects the increased mathematical sophistication of physics graduate students, and the fact that Lie groups have become as essential to modern theoretical physics as complex analysis and partial differential equations. Finite group theory -- the subject of Van's course -- is now summarily dealt with in the first seven pages of this volume.

This is not a book on modern mathematics. It deals with the application of rather old mathematics to very new physics. Most of the mathematical concepts were developed by Marius Sophus Lie and Elie-Joseph Cartan in the nineteenth century. Much of the physics is of the past decade. Here, the reader will find no narcissistic cry of mathematics for mathematics' sake. To the contrary, we physicist-chauvinist-pigs regard mathematics as the mare handmaiden of physics. Flights of mathematical fancy are tolerated only insofar as they are tethered to observable physical phenomena.

The use of continuous groups and their unitary representations is not something new in theoretical physics.

Spherical harmonics, after all, are nothing more than irreducible representations of the rotation group. Electron spin -- a mystery in its time -- simply shows that quantum mechanics (unlike classical mechanics) can gainfully employ the two-valued representations of the rotation group. Wigner has shown how mass and spin characterize the physically relevant representations of the Poincaré group. Heisenberg's isotopic spin can be understood most easily in terms of a group of unitary matrices acting on protons and neutrons. However, in all these examples, group theory, while elegant and pedagogically useful, is hardly essential. The basic concepts can be (and were) developed without explicit reference to the theory of Lie groups.

More recently, group theory has begun to play a central role. In the 1950s theoretical physicists searched for a "higher symmetry group" that would include both isotopic spin and hypercharge, and could organize the newly-discovered strange particles into meaningful patterns. The correct model [variously called "the eightfold way," "unitary spin," or "flavor SU(3)"] was invented in 1961 by Murray Gell-Mann and Yuval Ne'eman. It was not immediately and universally accepted. Some physicists doubted that particles so different as kaons and pions could be usefully regarded as members of the same supermultiplet. With the triumphant discovery of the predicted Ω^- particle in 1964, the doubters were silenced and it became clear that theoretical physicists had to understand Lie groups.

For a time, physicists were puzzled by the appearance of such a complicated group as SU(3) at a fundamental level of particle physics. So it was in an earlier era when crystallographers discovered the physical relevance of the 32 point groups. Just as crystal symmetries were eventually explained in terms of atomic theory, so were the successes of flavor SU(3) to be understood in terms of quarks. Nature did not, after all, use flavor SU(3) in any fundamental way. Its utility merely reflects the existence of three light quark species. With the discovery of the fourth and heavier charmed quark, some physicists were led astray by their mathematics. Finding SU(3) good, they could not resist flavor SU(4). Their predictions of the properties of charmed particles were wildly off the mark. Howard Georgi finds a moral here, which pervades much of his book: "Symmetry is a tool that should be used to determine the underlying dynamics, which must in turn explain the success (or failure) of the symmetry arguments. Group theory is a useful technique, but it is no substitute for physics."

A more fundamental role of Lie groups in physics relates to the recent and successful development of non-Abelian gauge theories of strong and electroweak interactions. These interactions result fron the exact invariance of the Lagrangian under a three-component Lie group. It appears as if this theory offers a complete and correct description of all (or perhaps, almost all) elementary-particle phenomena at accessible energies. The description of the three fundamental particle forces in terms of a three-component Lie group suggests a further unification in terms of a single simple Lie group. Howard and I invented the simplest such theory based on the group SU(5). Theories of this kind offer an explanation for the quantization of electric charge -- the fact that all observed electric charges are commensurate. Moreover, they predict the eventual, but observable, instability of all nuclear matter. The current literature is rife with even more elaborate unified theories based on larger Lie groups. The questions of the past can be asked anew: How can particles so different as quarks and leptons be placed in the same supermultiplet? Why does nature use a group as complicated as SU(5)? Will there be a new layer of structure below the now-fundamental quarks and leptons? We cannot yet answer these questions, but it is clear that a command of the simple and beautiful theory of Lie groups will be needed. For the future, we can only repeat the remarks of an earlier Harvard colleague, P.W. Bridgman, who wrote in 1927 that "whatever may be one's opinion as to the simplicity of either the laws or the material structure of nature, there can be no question that the possessors of some such conviction have a real advantage in the race for physical discovery. Doubtless there are many simple connections still to be discovered, and he who has a strong conviction of the existence of these connections is much more likely to find them than he who is not at all sure they are there." Amen.

Sheldon L. Glashow

PREFACE

Since 1976, I have taught a course in group theory in the physics department at Harvard. I wanted this course to reflect the practical knowledge of Lie algebras I had gleaned in struggling with unified gauge theories. But, I found that the textbooks on the subject either did not deal with the material I wanted to cover, or included so much material in such a compact mathematical language that I could not understand them myself. In the lecture notes on which this book is based, I tried to include all the simple ideas that I had actually found useful.

The book is intended as a textbook for beginning graduate students and advanced undergraduates, though I suspect that some practicing physicists may find it a useful reference. It should be accessible to students who have had a good undergraduate quantum mechanics course. I start with the operator treatment of angular momentum and build the theory of Lie algebras from there. The book is unique in that I make real use of both root-weight and tensor methods, going back and forth and applying insights from both approaches.

I illustrate most of the mathematical concepts with practical examples from modern particle physics. In addition to the standard examples of isospin, and approximate SU(3) and SU(6), I discuss color SU(3), heavy quarks, and the grand unified theories based on SU(5) and SO(10).

I should stress that this is a book about physics. Group theory is presented as a handy tool, to be used when it is appropriate to simplify the description of a physical system and, just as important, to be discarded when it is not appropriate. I have made this point very explicitly in Chapter XVII. I think that group theory, perhaps because it seems to give information for free, has been more misused in contemporary particle physics than any other branch of mathematics, except geometry. Students should learn the difference between physics and mathematics from the start.

I have tried to avoid mathematicisms whenever possible. But I do not think that I have been entirely successful. Sometimes the temptation to use a compact and elegant notation is too great. Partly for that reason, I have included many problems designed to liberate the student from the specific conventions used in the text.

My lecture notes were put into readable form by Paula Constantine. Her indefatigable efforts to make the manuscript presentable were an inspiration. I am also grateful to Marie Machacek for proofreading and helpful suggestions. Special thanks go to all the members of my family. They were very patient with me while I was putting the book together.

Finally, let me recount my first exposure to the FRONTIERS IN PHYSICS Series. It came in 1964 when I was a freshman at Harvard. I was bored stiff in a course on optics and modern physics when my section leader (a graduate student in experimental high energy physics) suggested that I pick up a little book by Richard Feynman called Theory of Fundamental Processes. It was my first look into the peculiar world of particle physics. I was hooked. I am pleased and honored to contribute to the series that got me started in this unique field.

Howard Georgi

LIE ALGEBRAS in PARTICLE PHYSICS

From Isospin to Unified Theories

I. GROUPS AND REPRESENTATIONS

A _group_ is a set G on which a multiplication operation $\cdot$ is defined with the following properties:

(I.1) If x and y are in G, $x \cdot y$ is in G;

(I.2) There is an identity element e in G such that $e \cdot x = x \cdot e = x$ for any x in G;

(I.3) For every x in G, there is an inverse element in G called x^{-1} such that $x \cdot x^{-1} = x^{-1} \cdot x = e$;

Howard Georgi, Lie Algebras in Particle Physics: From Isospin to Unified Theories

ISBN 0-8053-3153-0

(I.4) For every x, y and z in G, $(x \cdot y) \cdot z = x \cdot (y \cdot z)$.

For example, the integers form a group under addition, that is, the "multiplication" law $x \cdot y$ is defined as $(x+y)$. This is called the additive group of the integers.

Another example is the possible permutations of three objects. Suppose we have three things, labeled a, b and c, in three positions, labeled 1, 2 and 3. The elements of the permutation group are the six ways we can reshuffle the objects:

( ), do nothing, so that (a, b, c) remains (a, b, c);
(12), interchange the objects in positions 1 and 2, so (a, b, c) → (b, a, c);
(23), (a, b, c) → (a, c, b);
(13), (a, b, c) → (c, b, a);
(123), take the object in 1 and put it in two, etc. (a, b, c) → (c, a, b) (this is called a cyclic permutation);
(321), (a, b, c) → (b, c, a).

There is a natural multiplication law for such transformations of a physical system:

> $A \cdot B$ is the transformation obtained by first making the transformation B, then making the transformation A. (I.5)

For example,

$$(23)\cdot(12) = (321) \tag{I.6}$$

because

$$\begin{aligned} &(12) \text{ takes } (a, b, c) \rightarrow (b, a, c), \\ &(23) \text{ takes } (b, a, c) \rightarrow (b, c, a). \end{aligned} \tag{I.7}$$

Thus

$$(23)\cdot(12) \text{ takes } (a, b, c) \rightarrow (b, c, a) = (321). \tag{I.8}$$

The permutation group is an example of a <u>transformation group</u> on a physical system. It is clear that the reversible transformations on a system form a group under the multiplication law (I.5). Doing nothing is the identity transformation. Undoing a transformation is the inverse transformation. Associativity, (I.4), follows because at each step in the product $x\cdot y\cdot z$, there is a definite physical state of the system. This class of groups is very important because it includes any possible symmetry of a physical system.

In quantum mechanics, a transformation of the system is associated with a unitary operator in the Hilbert space (we will ignore the possibility of antiunitary operators, because they are irrelevant to the groups we will study in the following chapters). Thus, a transformation group of a quantum mechanical system is associated with a mapping of the group into a set of unitary operators. For each x in G there is a $D(x)$ which is a unitary operator. Furthermore, the mapping must preserve the multiplication law (to make physical sense). Thus,

$$D(x)D(y) = D(x\cdot y). \tag{I.9}$$

for all x and y in G.

A mapping which satisfies (I.9) is called a <u>representation</u> of the group G. In fact, a representation can involve nonunitary linear operators so long as they satisfy the multiplication law, (I.9).

For example, the mapping

$$D(n) = e^{in\theta} \tag{I.10}$$

is a representation of the additive group of the integers, because

$$e^{in\theta} e^{im\theta} = e^{i(n+m)\theta} . \tag{I.11}$$

The following mapping is a representation of the permutation group on three objects:

$$D(\ ) = \begin{vmatrix} 1 & 0 & 0 \\ 0 & 1 & 0 \\ 0 & 0 & 1 \end{vmatrix}, \quad D(12) = \begin{vmatrix} 0 & 1 & 0 \\ 1 & 0 & 0 \\ 0 & 0 & 1 \end{vmatrix},$$

$$D(13) = \begin{vmatrix} 0 & 0 & 1 \\ 0 & 1 & 0 \\ 1 & 0 & 0 \end{vmatrix}, \quad D(23) = \begin{vmatrix} 1 & 0 & 0 \\ 0 & 0 & 1 \\ 0 & 1 & 0 \end{vmatrix}, \tag{I.12}$$

$$D(123) = \begin{vmatrix} 0 & 1 & 0 \\ 0 & 0 & 1 \\ 1 & 0 & 0 \end{vmatrix}, \quad D(321) = \begin{vmatrix} 0 & 0 & 1 \\ 1 & 0 & 0 \\ 0 & 1 & 0 \end{vmatrix}.$$

The multiplication (12)(23) = (321) is mapped into the matrix multiplication

$$\begin{vmatrix} 0 & 1 & 0 \\ 1 & 0 & 0 \\ 0 & 0 & 1 \end{vmatrix} \begin{vmatrix} 1 & 0 & 0 \\ 0 & 0 & 1 \\ 0 & 1 & 0 \end{vmatrix} = \begin{vmatrix} 0 & 0 & 1 \\ 1 & 0 & 0 \\ 0 & 1 & 0 \end{vmatrix}. \tag{I.13}$$

In other words, a group is a multiplication table

satisfying (I.1-4). A representation is a specific realization of the multiplication by (finite or infinite dimensional) matrices. Group theory makes it possible to determine many properties of any representation from the abstract properties of the group.

Notice that the definition (I.1-4) does not require commutativity ($x \cdot y = y \cdot x$). In fact, while the additive group of the integers is commutative, the permutation group is not. A commutative group is called Abelian and a non-commutative group is non-Abelian. The permutations are non-Abelian, since for example (23)(12) = (123).

We will find it convenient to view representations both as abstract linear operators and as matrices. The connection is as follows: Let $|i\rangle$ be an orthonormal basis in the space on which $D(g)$ acts as a linear operator. Then

$$[D(g)]_{ij} = \langle i|D(g)|j\rangle. \tag{I.14}$$

So

$$D(g)|i\rangle = \sum_j |j\rangle\langle j|D(g)|i\rangle = |j\rangle[D(g)]_{ji}. \tag{I.15}$$

From now on, we will freely translate from one language to the other.

Two representations D_1 and D_2 are equivalent if they are related by a similarity transformation

$$D_2(x) = S\, D_1(x) S^{-1} \tag{I.16}$$

with a fixed operator S for all x in the group G.

A representation D is reducible if it is equivalent to a representation D' with block-diagonal form:

$$D'(x) = S\, D(x) S^{-1} = \begin{vmatrix} D_1'(x) & 0 \\ 0 & D_2'(x) \end{vmatrix}. \tag{I.17}$$

The vector space on which D' acts breaks up into two orthogonal subspaces, each of which is mapped into itself by all the operators $D'(x)$. The representation D' is said to be the direct sum of D'_1 and D'_2,

$$D' = D'_1 \oplus D'_2. \tag{I.18}$$

A representation is irreducible if it is not reducible, that is if it cannot be put into block diagonal form by a similarity transformation.

We will almost never talk about the group elements as abstract mathematical objects. Instead, we will think of them in one of their representations, as linear operators. This is good enough, since we can use a representation to obtain the multiplication table which is, in some abstract sense, the group. Furthermore, for the groups we will study, all the irreducible representations are equivalent to representations by unitary operators. Since it is the unitary representations we are interested in for quantum mechanics, we will therefore assume that we are always dealing with representations by unitary operators.

PROBLEMS FOR CHAPTER I

(I.A) If x is an element of a group G, prove that the inverse element x^{-1} is unique.

(I.B) Find the multiplication law for a group with three elements and prove that it is unique.

(I.C) Show that the representation (I.12) of the permutation group is reducible. Hint: find a vector which is an eigenvector of all the D's, then take it from there.

II. LIE GROUPS AND LIE ALGEBRAS

Compact Lie groups are groups of unitary operators in which the group elements are labeled by a set of continuous parameters. We will be interested in special groups called <u>compact Lie groups</u>. A Lie group is a group in which the elements are labeled by a set of continuous parameters with a multiplication law that depends smoothly on the parameters. The term "compact" refers to a global property of the group which we will not discuss in detail. In a

Howard Georgi, Lie Algebras in Particle Physics: From Isospin to Unified Theories ISBN 0-8053-3153-0

certain sense, the volume of the parameter space for a compact group is finite.

Any representation of a compact Lie group is equivalent to a representation by unitary operators. Any group element which can be obtained from the identity by continuous changes in the parameters can be written as

$$\exp\{i\,\alpha_a X_a\} \qquad \text{(II.1)}$$

where α_a (a=1 to N) are real parameters and X_a are linearly independent hermitian operators, and a sum over the repeated index a is implied. The set of all linear combinations $a_a X_a$ is a vector space, and X_a are a basis in the space. We will use the term <u>group generator</u> to refer (interchangably) to an arbitrary element of the vector space or specifically to the basis vectors X_a.

There are several spaces that are relevant here. Do not confuse the space of the group generators (which is N dimensional because $a = 1$ to N) with the space on which the generators act, which is some as yet unspecified Hilbert space. In fact, for the compact Lie groups, we can always take the space on which the generators act to be finite dimensional, so that you can think of the X_a as finite hermitian matrices.

There are two nice things about the generators. The first is simply that they form a vector space. Unlike group elements themselves, the generators can be multiplied by numbers and added to obtain other generators. The second is that they satisfy simple commutation relations which determine (almost) the full structure of the group. Consider the product

$$\begin{aligned}&\exp\{i\lambda X_b\}\exp\{i\lambda X_a\}\exp\{-i\lambda X_b\}\exp\{-i\lambda X_a\}\\ &\quad = 1+\lambda^2[X_a, X_b]+\cdots\end{aligned}$$

Because of the group property, the product of group elements is another group element and can be written as $\exp\{i\,\beta_c X_c\}$. As $\lambda \to 0$ we must have

$$\lambda^2 [X_a, X_b] \to i\,\beta_c X_c$$

$$\beta_c = \lambda^2 f_{abc} \quad \text{so}$$

$$[X_a, X_b] = i\, f_{abc} X_c . \tag{II.2}$$

The constants f_{abc} are called the structure constants of the group. Clearly they are determined by the group multiplication law. They also determine the multiplication law (at least for elements continuously connected to the identity) as follows:

$$\exp\{i\,\alpha_a X_a\}\exp\{i\,\beta_b X_b\} \equiv \exp\{i\,\delta_c X_c\} \tag{II.3}$$

where δ_c is determined by α, β and f

$$\delta_c = \alpha_c + \beta_c - \frac{1}{2} f_{abc}\,\alpha_a \beta_b + \cdots \tag{II.4}$$

This identity follows in a straightforward way from the definition of $\exp\{A\}$. We can obtain δ_a to any desired order in α_a and β_a just by expanding both sides of (II.3) and comparing.

The generators also satisfy the following identity:

$$[X_a, [X_b, X_c]] + \text{cyclic permutations} = 0. \tag{II.5a}$$

This is called the Jacobi identity. It is obvious for a representation, since then the X_a are just linear operators, but in fact it is true for the abstract group generators. In terms of the structure constants, (II.5a) becomes

$$f_{bcd} f_{ade} + f_{abd} f_{cde} + f_{cad} f_{bde} = 0, \tag{II.5b}$$

If we define a set of matrices T_a

$$(T_a)_{bc} \equiv -i\, f_{abc} \tag{II.6}$$

then (II.5) can be rewritten as

$$[T_a, T_b] = i f_{abc} T_c.$$

In other words, the structure constants themselves generate a representation of the algebra. The representation generated by the structure functions is called the _adjoint representation_. The _dimension_ of a representation is the dimension of the vector space on which it acts. The dimension of the adjoint representation is just the number of generators, which is the number of real parameters necessary to describe a group element.

The generators and the commutation relations define the _Lie algebra_ associated with the Lie group. Clearly every representation of the group defines a representation of the algebra. The generators in the representation, when exponentiated, give the operators of the group representation.

The definitions of equivalence, reducibility and irreducibility can be transferred from the group to the algebra with no change.

The structure constants depend on what basis we choose in the vector space of the generators. To choose an appropriate basis, we use the adjoint representation, (II.6). Consider the trace $\mathrm{tr}(T_a T_b)$. This is a real symmetric matrix, so we can diagonalize it by choosing appropriate real linear combinations of the X^a's and therefore of the T^a's. Suppose we have done this, so that

$$\mathrm{tr}(T_a T_b) = k_a \delta_{ab} (\text{no sum}). \tag{II.7}$$

We still have the freedom to rescale the generators, so for example, we could choose all the nonzero k^a's to have absolute value 1. But, we cannot change the sign of the k^a's.

In this book, we will be concerned almost entirely

with algebras in which all the k^a's are positive. For these algebras, which are associated with the <u>compact semi-simple</u> Lie groups, we can take

$$tr(T_a T_b) = \lambda \, \delta_{ab} \tag{II.8}$$

for any convenient positive λ. In this basis, the structure constants are completely antisymmetric, because we can write $f_{abc} = -i \, \lambda^{-1} \, tr([T_a, T_b]T_c)$, which is completely antisymmetric because of the cyclic property of the trace.

Notice that in this basis, the generators in the adjoint representation are hermitian matrices. In fact, one can show that for the compact Lie groups (as for finite groups) any representation is equivalent to a representation by hermitian operators and all irreducible representations are finite hermitian matrices.

A few more definitions may help the reader to appreciate when the structure constants can be useful. An <u>invariant subalgebra</u> is some set of generators which goes into itself (or zero) under commutation with any element of the algebra. That is if X is any generator in the invariant subalgebra and Y is any generator in the whole algebra, [X, Y] is a generator in the invariant subalgebra (or it is zero).

An algebra which has no nontrivial invariant subalgebra (the whole algebra, and the set of no generators at all are always trivial invariant subalgebras) is called <u>simple</u>. A simple algebra generates a simple group.

Particularly annoying are Abelian invariant subalgebras. The generators in an Abelian invariant subalgebra commute with everything. Each of these generators is associated with what we will call a <u>U(1)</u> factor of the group. U(1) is the group of phase transformations. At any rate, U(1) factors do not show up in the structure constants at all. If X_a is a U(1) generator, $f_{abc} = 0$ for all b and c and $k_a = 0$. The structure constants don't tell us anything about the Abelian invariant subalgebras.

Algebras without Abelian invariant subalgebras are called <u>semisimple</u>. They are built, as we will see, by putting simple algebras together. In these algebras, every generator has a nonzero commutator with some other generator. In such a case the structure constants carry a great deal of information. We will use them to determine the entire structure of the algebra and its representations. From here on, unless explicitly stated, we will be discussing semisimple algebras, and we will deal with representations by unitary operators.

The generators (like the linear operators of the representations they generate) can be thought of as either linear operators or matrices via

$$X_a|i\rangle = |j\rangle\langle j|X_a|i\rangle = |j\rangle[X_a]_{ji}. \qquad \text{(II.9)}$$

Note! In a matrix language, the states form row vectors and the matrix representing a linear operator acts on the right.

In the Hilbert space on which a representation acts, the group elements can be thought of as a transformation on the states. The group element $\exp\{i\alpha_a X_a\}$ maps or "transforms" the kets as follows:

$$|i\rangle \to \exp\{i\alpha_a X_a\}|i\rangle. \qquad \text{(II.10)}$$

Then the corresponding bras transform as

$$\langle i| \to \langle i|\exp\{-i\alpha_a X_a\}. \qquad \text{(II.11)}$$

The ket obtained by acting on $|i\rangle$ with an operator O must also transform as in (II.10). This implies that any operator O transforms as follows:

$$O \rightarrow \exp\{i\alpha_a X_a\} O \exp\{-i\alpha_a X_a\}. \qquad (II.12)$$

(II.10-12) imply that all matrix elements are unchanged under the transformation.

To obtain the action of the algebra on these objects, we differentiate with respect to α and divide by i. Thus, corresponding to the action of the generator X_a on a ket

$$X_a|i\rangle, \qquad (II.13)$$

is $-X_a$ acting on a bra

$$-\langle i|X_a, \qquad (II.14)$$

and the commutator of X_a with an operator

$$[X_a, O]. \qquad (II.15)$$

Then the invariance of a matrix element $\langle i|O|j\rangle$ is expressed by the statement,

$$\langle i|O(X_a|j\rangle) + \langle i|[X_a, O]|j\rangle + (-\langle i|X_a)O|j\rangle = 0. \qquad (II.16)$$

PROBLEMS FOR CHAPTER II

(II.A) Find all components of the matrix $\exp\{i\alpha A\}$, where

$$A = \begin{pmatrix} 0 & 0 & 1 \\ 0 & 0 & 0 \\ 1 & 0 & 0 \end{pmatrix}.$$

(II.B) If $[A, B] = B$, calculate $\exp\{i\alpha A\}B\exp\{-i\alpha A\}$.

(II.C) Carry out the expansion of δ_c in (II.4) to third order in α and β.

(II.D) Prove Schur's Lemma, which states that if a matrix M commutes with all the matrix generators of an irreducible representation of a Lie algebra, M is proportional to the identity matrix. Hint: assume M is hermitian and diagonalize it.

III. SU(2)

The simplest non-Abelian Lie algebra consists of three generators J_a, $a = 1$ to 3 satisfying $f_{abc} = \varepsilon_{abc}$, where ε_{abc} is the completely antisymmetric tensor ($\varepsilon_{123} = 1$ and ε_{abc} is antisymmetric in exchange of any two indices). The commutation relations should be familiar:

$$[J_1, J_2] = i\, J_3, \quad [J_2, J_3] = i\, J_1$$

$$[J_3, J_1] = i\, J_2. \tag{III.1}$$

Howard Georgi, Lie Algebras in Particle Physics: From Isospin to Unified Theories ISBN 0-8053-3153-0

This is the angular momentum algebra. As is well known, the generators of the rotation group satisfy (III.1). We will analyze the representations of the algebra (III.1) in two different ways, both of which generalize to other Lie groups.

The first technique is essentially the operator method which I hope is familiar from introductory quantum mechanics courses. But, we will do it in a purposely cumbersome way, without using the special properties of the operator $J_a J_a$.

Assume we are given an N dimensional irreducible representation of (III.1). We can diagonalize the hermitian operator J_3. Consider the states with the highest value of J_3. If there is more than 1, label them by integers $\alpha = 1, 2 \cdots$.

$$J_3 |j, \alpha\rangle = j |j, \alpha\rangle \qquad \text{(III.2a)}$$

Choose $|j, \alpha\rangle$ such that

$$\langle j, \alpha | j, \beta\rangle = \delta_{\alpha\beta}. \qquad \text{(III.2b)}$$

Define the raising and lowering operators

$$J^{\pm} = (J_1 \pm i J_2)/\sqrt{2} \qquad \text{(III.3)}$$

so called because when they act on an eigenstate of J_3 they raise or lower the eigenvalue by one unit. The commutation relations for these are

$$[J_3, J^{\pm}] = \pm J^{\pm} \tag{III.4a}$$

$$[J^{+}, J^{-}] = J_3 \tag{III.4b}$$

so if $J_3|m\rangle = m|m\rangle$, then

$$J_3J^{\pm}|m\rangle = J^{\pm}J_3|m\rangle \pm J^{\pm}|m\rangle$$
$$= (m \pm 1)J^{\pm}|m\rangle.$$

Since j is the highest J_3 eigenvalue, we must have $J^{+}|j, \alpha\rangle = 0.$

The states $J^{-}|j, \alpha\rangle$ have $J_3 = j - 1$. Define

$$J^{-}|j, \alpha\rangle \equiv N_j(\alpha)|j - 1, \alpha\rangle, \tag{III.6}$$

States with different α are orthogonal because

$$N_j^{*}(\beta)N_j(\alpha)\langle j - 1, \beta|j - 1, \alpha\rangle$$
$$= \langle j, \beta|J^{+}J^{-}|j, \alpha\rangle$$
$$= \langle j, \beta|[J^{+}, J^{-}]|j, \alpha\rangle$$
$$= \langle j, \beta|J_3|j, \alpha\rangle = j\langle j, \beta|j, \alpha\rangle$$
$$= j\ \delta_{\alpha\beta}.$$

So we can choose $|j - 1, \alpha\rangle$ orthonormal, then $N_j(\alpha)^2 = j$. Clearly we can choose phases so $N_j(\alpha) = \sqrt{j} = N_j$. Then,

$$J^{+}|j - 1, \alpha\rangle = \frac{1}{N_j} J^{+}J^{-}|j, \alpha\rangle$$
$$= \frac{1}{N_j}[J^{+}, J^{-}]|j, \alpha\rangle = N_j|j, \alpha\rangle.$$

Note that $J^{\pm}$ change j <u>without changing α</u>.

Now we can look at $J^{-}|j - 1, \alpha\rangle$ and show by an analogous argument that there are orthonormal states $|j - 2, \alpha\rangle$ satisfying

$$J^-|j-1,\alpha\rangle = N_{j-1}|j-2,\alpha\rangle,$$
$$J^+|j-2,\alpha\rangle = N_{j-1}|j-1,\alpha\rangle. \tag{III.7}$$

Continuing the process we find orthonormal states $|j-k,\alpha\rangle$, such that

$$J^-|j-k,\alpha\rangle = N_{j-k}|j-k-1,\alpha\rangle,$$
$$J^+|j-k-1,\alpha\rangle = N_{j-k}|j-k,\alpha\rangle. \tag{III.8}$$

The N's (chosen real) satisfy

$$\begin{aligned} N^2_{j-k} &= \langle j-k,\alpha|J^+J^-|j-k,\alpha\rangle \\ &= \langle j-k,\alpha|[J^+,J^-]|j-k,\alpha\rangle \\ &\quad + \langle j-k,\alpha|J^-J^+|j-k,\alpha\rangle \\ &= N^2_{j-k+1} + j - k. \end{aligned} \tag{III.9}$$

This recursion relation for N_{j-k} can be solved by addition:

$$\begin{aligned} N^2_j &= j \\ N^2_{j-1} - N^2_j &= j - 1 \\ &\vdots \\ N^2_{j-k} - N^2_{j-k+1} &= j - k, \\ \hline N^2_{j-k} &= (k+1)j - k(k+1)/2 \\ &= \frac{1}{2}(k+1)(2j-k), \end{aligned}$$

or with $j - k = m$,

$$N_m = \frac{1}{\sqrt{2}}\sqrt{(j+m)(j-m+1)}\,. \tag{III.10}$$

Eventually, we must come to a $j-\ell$ such that $J^-|j-\ell,\alpha\rangle = 0$. But this requires $N_{j-\ell} = 0$ or $\ell = 2j$.

So, $j = \ell/2$ for some integer ℓ.

The space breaks up into α subspaces of dimension

2j + 1 and the subspace of states which we have not already found by our procedure. But the J's act on these subspaces in block diagonal form

$$\begin{vmatrix} J & 0 & 0 & 0 & 0 \\ 0 & J & 0 & 0 & 0 \\ 0 & 0 & J & 0 & 0 \\ 0 & 0 & 0 & J & 0 \\ 0 & 0 & 0 & 0 & ? \end{vmatrix} \quad \begin{matrix} \alpha = 1 \\ \alpha = 2 \\ \alpha = 3 \\ \text{etc.} \\ \text{others} \end{matrix} \tag{III.11}$$

Since we assumed the representation was irreducible, there must be only one α and no other states.

We have explicitly constructed the irreducible representations of (III.1). They are characterized by their highest J_3 eigenvalue j which must be an integer or half-odd-integer. They are 2j + 1 dimensional.

Furthermore, the same construction can be used to decompose an arbitrary representation into its irreducible components. Starting with any representation, we obtain (III.11) in which the irreducible representations with the highest j are explicitly separated out. Now we can apply the procedure again to the other states to find the representations with the next highest j, and so on until the representation is completely reduced.

The values of J_3 are called _weights_, and j is the _highest weight_. This construction, which characterizes the irreducible representations in terms of their highest weight, generalizes to any Lie algebra.

The representation of (III.1) with highest weight j is sometimes called the _spin j representation_ because it is the representation associated with a particle at rest with spin angular momentum $j(J^2 = j(j + 1)\hbar^2)$.

When hermitian angular momentum operators (that is J's satisfying (III.1)) exist, the entire Hilbert space is a representation and can be decomposed into states labeled $|m, j, \alpha\rangle$ in the spin j representation with $J_3 = m$. The α stands for all other labels necessary to characterize the states. We can (and almost always will) choose the state orthonormal in α, so the states satisfy

$$\langle m, j, \alpha | m', j', \beta\rangle = \delta_{mm'} \delta_{jj'} \delta_{\alpha\beta}. \tag{III.12}$$

These states form a convenient basis for studying the properties of the system under rotations.

The δ functions on the right hand side of (III.12) are not purely a convention. They follow from the algebraic structure. We know that m and m' must be equal for the matrix element to be nonzero because they are eigenvalues of a hermitian operator. Of course, we know from the familiar analysis in terms of the operator J^2 that j and j' are also eigenvalues of a hermitian operator. But we can also show that $j = j'$ for a nonzero matrix element directly with raising and lowering operators. Suppose $j' > j$ and consider

$$\begin{aligned} &\langle j, j, \alpha | j, j', \beta\rangle \\ &= \left|\frac{2}{(j' + j)(j' - j + 1)}\right|^{1/2} \langle j, j, \alpha | J^- | j + 1, j', \beta\rangle \\ &= 0. \end{aligned}$$

Applying $J^{\pm}$ judiciously, we can show that $j = j'$ for all m.

The highest weight construction shows us how to break down a representation into simpler irreducible representations. There is another way of looking at it. We can build up an arbitrary representation out of the simplest representation. The simplest nontrivial representation of (III.1) is generated by $\sigma_a/2$ where σ_a are the Pauli matrices:

$$\sigma_1 = \begin{vmatrix} 0 & 1 \\ 1 & 0 \end{vmatrix}, \quad \sigma_2 = \begin{vmatrix} 0 & -i \\ i & 0 \end{vmatrix}, \quad \sigma_3 = \begin{vmatrix} 1 & 0 \\ 0 & -1 \end{vmatrix}. \qquad \text{(III.13)}$$

Notice that the group elements, $\exp\{i\,\alpha_a\sigma_a/2\}$, are the <u>S</u>pecial (determinant 1) <u>U</u>nitary <u>2 x 2</u> matrices. The group is often called SU(2).

Given two representations D_1(dim n) and D_2 (dim m) of a group G, we can form another representation in two ways. The direct sum $D_1 \oplus D_2$ (we have already discussed) has dimension n + m. It is formed by the block-diagonal matrices

$$\begin{vmatrix} D_1(g) & 0 \\ 0 & D_2(g) \end{vmatrix} \qquad \text{(III.14)}$$

The generators of the representation $D_1 \oplus D_2$ are

$$X_a^{D_1 \oplus D_2} = \begin{vmatrix} X_a^{D_1} & 0 \\ 0 & X_a^{D_2} \end{vmatrix}. \qquad \text{(III.15)}$$

The operation of taking the direct sum is the reverse of the process of reducing a representation.

We can also form a representation of dimension n x m as follows. If $|i\rangle$ i = 1 to n is an orthonormal basis in the space on which D_1 acts and $|x\rangle$ x = 1 to m is orthonormal basis for D_2, we can identify the product $\{|i\rangle|x\rangle\}$ with an orthonormal basis in an n x m dimensional space, called the <u>direct product space</u>. On this space, the <u>direct product representation</u> $D_1 \otimes D_2$ is

$$(D_1 \otimes D_2)(g)\{|i\rangle|x\rangle\} = \{D_1(g)|i\rangle D_2(g)|x\rangle\}.$$

In the matrix language

$$[(D_1 \otimes D_2)(g)]_{ix,jy} = [D_1(g)]_{ij}[D_2(g)]_{xy}. \quad \text{(III.16)}$$

The generators of the direct product representation are the sums (obtained by differentiating the product in (III.16))

$$[X_a^{D_1 \otimes D_2}]_{ix,jy} = [X_a^{D_1}]_{ij}\ \delta_{xy} + \delta_{ij}[X_a^{D_2}]_{xy}. \quad \text{(III.17)}$$

Clearly we can form the direct product of an arbitrary number of representations.

The ideas of direct product space and direct product representation are very important. The space occurs whenever a state can be labeled by two or more independent indices. When the indices are associated with representations of a Lie algebra, the states are in a direct product representation. The addition of angular momentum in quantum mechanics is an example of (III.17). Often, the δ functions in (III.17) are suppressed, since they are just identity matrices in the relevant space. The important thing to remember in dealing with direct products is what matrices act on which space.

Now getting back to the two dimensional representation of SU(2), we can form the direct product of n of them

$$[D \otimes \cdots \otimes D]_{i_1 \cdots i_n,\ j_i \cdots j_n} = D_{i_1 j_1} \cdots D_{i_n j_n}$$

(III.18)

acting on the n component objects $u_{j_1 \cdots j_n}$. This representation is reducible. In particular, (III.18) preserves any symmetry of the u vectors under permutation of the labels 1 to n. The product representation breaks up into irreducible components corresponding to vectors with definite symmetry. The component acting on the completely symmetric subspace is n + 1 dimensional, because the completely symmetric n index object has only n + 1 independent components. In fact, this representation is the irreducible representation with $j = n/2$. This construction also generalizes to more complicated groups and their algebras. It is the basis of what are called "tensor methods" for dealing with representations.

PROBLEMS FOR CHAPTER III

(III.A) Use the highest weight decomposition to show that

$$\{j\} \otimes \{s\} = \sum_{\oplus |s-j|}^{s+j} \{\ell\}$$

where $\{j\}$ is the spin j representation of SU(2).

(III.B) Calculate $\exp\{i\vec{r}\cdot\vec{\sigma}\}$ where $\vec{\sigma}$ are the Pauli matrices. Hint: write $\vec{r} = |\vec{r}| \cdot \hat{r}$.

(III.C) Show explicitly that the spin 1 representation obtained by the highest weight procedure, (III.8 and 10) with j=1, is equivalent to the adjoint representation, (II.6) with $f_{abc} = \varepsilon_{abc}$.

(III.D) Suppose that $(\sigma_a)_{ij}$ and $(\eta_a)_{xy}$ are Pauli matrices in two different two dimensional spaces. In the four dimensional tensor product space define the basis

$|1\rangle = |i=1\rangle|x=1\rangle$, $|2\rangle = |i=1\rangle|x=2\rangle$,

$|3\rangle = |i=2\rangle|x=1\rangle$ $|3\rangle = |i=2\rangle|x=2\rangle$.

Write out the matrix elements of $\sigma_2 \otimes \eta_1$ in this basis.

(III.E) We will often abbreviate the tensor product notation by leaving out the indices and the identity matrices. This makes for very compact equations, but you must keep your wits about you to stay in the right space. In the example of problem (III.D), we could write:

$(\sigma_a)_{ij}(\eta_b)_{xy}$ as $\sigma_a\eta_b$;

$(\sigma_a)_{ij}\delta_{xy}$ as σ_a;

$\delta_{ij}(\eta_a)_{xy}$ as η_a;

$\delta_{ij}\delta_{xy}$ as 1.

To get some practice with this notation, calculate

(a) $[\sigma_a, \sigma_b\eta_c]$,

(b) $\mathrm{tr}(\sigma_a\{\eta_b, \sigma_c\eta_d\})$,

(c) $[\sigma_1\eta_1, \sigma_2\eta_2]$.

Remember that σ_a and η_b commute.

IV. TENSOR OPERATORS AND THE WIGNER-ECKART THEOREM

A <u>tensor operator</u> O_x (x=1 to 2s+1) is a set of linear operators with simple commutation relations with the angular momentum operators:

$$[J_a, O_x] = O_y [J_a^s]_{yx} \tag{IV.1}$$

where J_a^s are the matrix generators of the irreducible spin s representation. The operators O_x are said to transform

Howard Georgi, Lie Algebras in Particle Physics: From Isospin to Unified Theories ISBN 0-8053-3153-0

according to the spin s representation.

For example, consider a system of a particle in a spherically symmetric potential. Here J_a are the orbital angular momentum operators

$$J_a = L_a = \varepsilon_{abc} r_b p_c . \tag{IV.2}$$

The position vector r_a is a tensor operator because

$$\begin{aligned}[J_a, r_b] &= \varepsilon_{acd}[r_c p_d, r_b] \\ &= -i\,\varepsilon_{acb} r_c = r_c [J_a^1]_{cb} .\end{aligned} \tag{IV.3}$$

It transforms according to the spin 1 (adjoint) representation.

The reason that the definition (IV.1) is useful is this: Consider the state $O_x|m, j, \alpha>$. It satisfies

$$\begin{aligned}&J_a\{O_x|m, j, \alpha>\} \\ &\quad = [J_a, O_x]|m, j, \alpha> + O_x J_a|m, j, \alpha> \\ &\quad = O_y|m, j, \alpha>[J_a^s]_{yx} \\ &\quad + O_x|m', j, \alpha>[J_a^j]_{m'm} .\end{aligned} \tag{IV.4}$$

Comparing (IV.4) with (III.17), we see that the state $O_x|m, j, \alpha>$ transforms according to the tensor product representation $s \otimes j$.

We can choose linear combinations of the O_x as

follows, to diagonalize the hermitian matrix $[J_3^s]$:

$$O'_\ell = O_y [S^{-1}]_{y\ell} \qquad (\ell = -s \text{ to } s)$$

$$[S\, J_3^s\, S^{-1}]_{\ell\ell'} = \ell\, \delta_{\ell\ell'}. \tag{IV.5}$$

Then

$$[J_3, O'_\ell] = \ell\, O'_\ell \quad \text{and}$$

$$J_3 O'_\ell |m, j, \alpha\rangle = (\ell+m) O'_\ell |m, j, \alpha\rangle \tag{IV.6}$$

In the basis defined by (IV.5), the operators O'_ℓ are analogous to the J_3 eigenstates $|\ell, s\rangle$. The state $O'_\ell |m, j, \alpha\rangle$ is the analog of the tensor product state $\{|\ell, s\rangle |m, j, \alpha\rangle\}$. We can decompose it, just as we can decompose the product state into states which transform irreducibly under the algebra. You have shown in the exercises that $s \otimes j = (s+j) \oplus \cdots \oplus |s-j|$. Thus,

$$O'_\ell |m, j, \alpha\rangle = \sum_{J=|s-j|}^{s+j} \langle \ell+m, J | \ell, s; m, j\rangle \cdot |\ell+m, J\rangle. \tag{IV.7}$$

The constants $\langle \ell+m, J | \ell, s; m, j\rangle$ are pure group theory. They are called <u>Clebsch-Gordan coefficients.</u> You may be familiar with them from your study of addition of angular momenta in quantum mechanics. You can look them up in tables. Or better still, you can determine them yourself by working out the corresponding decomposition of the tensor product state

$$|\ell, s\rangle |m, j\rangle = \sum_{J=|s-j|}^{s+j} \langle \ell+m, J | \ell, s; m, j\rangle \cdot |\ell+m, J\rangle, \tag{IV.8}$$

using the highest weight decomposition algorithm. For example, we know that the state of highest weight is

$$|s, s\rangle|j, j\rangle = |s+j, s+j\rangle. \qquad \text{(IV.9)}$$

Thus

$$\langle s+j, s+j|s, s; j, j\rangle = 1. \qquad \text{(IV.10)}$$

Of course, it is equally obvious that the higest weight state in the product $O'_\ell|m, j, \alpha\rangle$ is $O'_s|j, j, \alpha\rangle$. But to go beyond this, it is much more convenient to work with analog tensor product states. Acting on (IV.9) with the lowering operator J^- we find

$$\sqrt{\frac{s}{s+j}}\,|s-1, s\rangle|j, j\rangle + \sqrt{\frac{j}{s+j}}\,|s, s\rangle|j-1, j\rangle = |s+j-1, s+j\rangle. \qquad \text{(IV.11)}$$

Thus (for example)

$$\langle s+j-1, s+j|s-1, s; j, j\rangle = \sqrt{\frac{s}{s+j}}\,. \qquad \text{(IV.12)}$$

Proceeding in this way we can find all of the coefficients for $J = s+j$. To get them for $J = s+j-1$, we proceed with the highest weight algorithm and throw out all the $J = s+j$ states. The highest weight state remaining is the state orthogonal to (IV.11):

$$\sqrt{\frac{j}{s+j}}\,|s-1, s\rangle|j, j\rangle - \sqrt{\frac{s}{s+j}}\,|s, s\rangle|j-1, j\rangle = |s+j-1, s+j-1\rangle. \qquad \text{(IV.13)}$$

Thus (for example)

$$\langle s+j-1, s+j-1|s, s; j-1, j\rangle = -\sqrt{\frac{s}{s+j}}\,. \qquad \text{(IV.14)}$$

(IV.13) is a definition. The state has been properly

normalized, but its phase is ours to choose. The Clebsch-Gordan coefficient (IV.14) depends on the choice of the phase. When you use tables of Clebsch-Gordan coefficients, you must be sure that you understand their phase convention. I always find it easier to work them out. Then I know what my phase convention is and as long as I use it consistently, I do not get into trouble.

Now we know by analogy that, for example, the state

$$\sqrt{\frac{j}{s+j}}\; O'_{s-1}|j, j, \alpha\rangle$$

$$-\sqrt{\frac{j}{s+j}}\; O'_{s}|j-1, j, \alpha\rangle$$

$$= |s+j-1, s+j-1\rangle \qquad \text{(IV.15)}$$

is the highest weight state in the $J = s+j-1$ representation. We can confirm this by applying J^+ to (IV.15) and checking that it annihilates the state. Indeed this must happen because the action of J^+ on (IV.15) is the same as the action of J^+ on (IV.13).

Something slightly nontrivial has happened here. We used the orthogonality property of the tensor product

$$(\langle \ell', s|\langle m', j|)(|\ell, s\rangle|m, j\rangle) = \delta_{\ell\ell'}\delta_{mm'} \qquad \text{(IV.16)}$$

to derive the state (IV.13) which is annihilated by J^+. We then argued that the analog state (IV.15) is also annihilated by J^+. And this is true even though the orthogonality property analogous to (IV.16) is not true:

$$\langle m', j, \alpha|O'^{\dagger}_{\ell'}\, O'_{\ell}|m, j, \alpha\rangle \neq \delta_{m'm}\delta_{\ell'\ell}. \qquad \text{(IV.17)}$$

Indeed, group theory doesn't tell us anything about how the state (IV.15) is normalized.

Now, back to (IV.7). We can decompose the states $|\ell+m, j\rangle$ which appear on the right-hand side of (IV.7) in terms of our complete set of states $|m, j, \alpha>$ as follows:

$$|\ell+m, J> = \sum_\beta k_{\beta\alpha}|\ell+m, J, \beta>. \qquad (IV.18)$$

The constants $k_{\beta\alpha}$ depend on j, α, and on the operator 0 which appear on the left-hand side of (IV.7) and also on J and β. <u>But $k_{\beta\alpha}$ is independent of ℓ and m</u>. It is enough to find $k_{\beta\alpha}$ for $\ell+m = J$ (say). This is obvious because $J^{\pm}$ applied to (IV.18) changes $\ell+m$ without changing the constants $k_{\beta\alpha}$. The $k_{\beta\alpha}$ are called <u>reduced matrix elements</u> and denoted by

$$k_{\beta\alpha} = <J, \beta|0|j, \alpha>. \qquad (IV.19)$$

Finally, we can put all this together to get what is called the <u>Wigner-Eckart theorem</u> for matrix elements of tensor operators:

$$<m', J, \beta|0'_\ell|m, j, \alpha>$$

$$= \delta_{m',\ell+m}<\ell+m, J|\ell, s; m, j>$$

$$\cdot <J, \beta|0|j, \alpha>. \qquad (IV.20)$$

According to this theorem, if we know any nonzero matrix element of a tensor operator between states of some given J, β and j, α, we can compute all the others using group theory. This sounds remarkable, but remember we just arrived at it using nothing more than the properties of the algebra. The point is simply that we can use $J^{\pm}$ to go up and down within representations using pure group theory. Thus, it should not be surprising that by judicious use of the raising and lowering operators we can relate the various matrix elements.

Let us work out an example involving the position operator. Given

$$<\frac{1}{2}, \frac{1}{2}, \alpha|r_3|\frac{1}{2}, \frac{1}{2}, \beta> = A, \tag{IV.21}$$

we should be able to find

$$<\frac{1}{2}, \frac{1}{2}, \alpha|r_1|-\frac{1}{2}, \frac{1}{2}, \beta>. \tag{IV.22}$$

The first step is to construct the operators r'_ℓ. This can be done easily by noting that $[J_3, r_3] = 0$, so that we can take

$$r'_0 = r_3. \tag{IV.23a}$$

Then the commutation relations for the spin 1 operator give the rest:

$$\begin{aligned}[J^\pm, r'_0] &= r'_{\pm 1} \\ &= \mp (r_1 \pm ir_2)/\sqrt{2}. \end{aligned} \tag{IV.23b}$$

Thus

$$A = <\frac{1}{2}, \frac{1}{2}, \alpha|r'_0|\frac{1}{2}, \frac{1}{2}, \beta>, \tag{IV.24}$$

and

$$\begin{aligned}&<\frac{1}{2}, \frac{1}{2}, \alpha|r_1|-\frac{1}{2}, \frac{1}{2}, \beta> \\ &= <\frac{1}{2}, \frac{1}{2}, \alpha\left|\frac{1}{\sqrt{2}}(-r'_{+1} + r'_{-1})\right|-\frac{1}{2}, \frac{1}{2}, \beta> \\ &= -\frac{1}{\sqrt{2}}<\frac{1}{2}, \frac{1}{2}, \alpha|r'_{+1}|-\frac{1}{2}, \frac{1}{2}, \beta>.\end{aligned}$$

Then the theorem (IV.20) gives

$$A = <\frac{1}{2}, \frac{1}{2}|0, 1; \frac{1}{2}, \frac{1}{2}><\frac{1}{2}, \alpha|r|\frac{1}{2}, \beta>, \tag{IV.26}$$

and the matrix element we want is

$$-\frac{1}{\sqrt{2}}<\frac{1}{2}, \frac{1}{2}|1, 1; -\frac{1}{2}, \frac{1}{2}><\frac{1}{2}, \alpha|r|\frac{1}{2}, \beta>. \tag{IV.27}$$

From (IV.13 and 14) (or your favorite table) we see that

$$\langle\tfrac{1}{2}, \tfrac{1}{2}|1, 1; -\tfrac{1}{2}, \tfrac{1}{2}\rangle = -\sqrt{2/3}$$

$$\langle\tfrac{1}{2}, \tfrac{1}{2}|0, 1; \tfrac{1}{2}, \tfrac{1}{2}\rangle = \sqrt{1/3}. \qquad \text{(IV.28)}$$

Thus

$$\langle\tfrac{1}{2}, \tfrac{1}{2}, \alpha|r_1|-\tfrac{1}{2}, \tfrac{1}{2}, \beta\rangle = A. \qquad \text{(IV.29)}$$

PROBLEMS FOR CHAPTER IV

(IV.A) Consider an operator O_x, x=1, 2, transforming according to the spin 1/2 representation, as follows

$$[J_a, O_x] = O_y \sigma^a_{yx}/2$$

where σ^a are the Pauli matrices, (III.13). Given

$$<-\frac{1}{2}, \frac{3}{2}, \alpha|O_1|-1, 1, \beta> = A,$$

find

$$<-\frac{3}{2}, \frac{3}{2}, \alpha|O_2|-1, 1, \beta>.$$

(IV.B) The operator $(r'_{+1})^2$ satisfies $[L^+, (r'_{+1})^2] = 0$. It is therefore the O_{+2} component of a spin 2 tensor operator. Construct the other components, O_m [Note that the product of tensor operators transforms like the tensor product of their representations]. What is their connection with the spherical harmonics, $Y_{\ell,m}(\theta, \phi)$. [Hint: let $r_1 = \sin\theta\cos\phi$, $r_2 = \sin\theta\sin\phi$, $r_3 = \cos\theta$]? Can you generalize this construction to arbitrary ℓ and explain what is going on?

V. ISOSPIN

We could fill a course with the study of the angular momentum algebra and the rotation group. Spacetime symmetries like rotations are particularly obvious examples of groups, because they are transformation groups with the multiplication law (I.5). There are other important examples: the Lorentz group of special relativity; and the Poincare group (the Lorentz group plus translations in space and time.) These, however, are not compact groups. The nature of their representations is different, and we will not discuss them here.

Howard Georgi, Lie Algebras in Particle Physics: From Isospin to Unified Theories ISBN 0-8053-3153-0

Instead, we will be concerned with groups which involve changes of particle identities, with no connection to the structure of space and time. These groups are called internal symmetries. The archetype is the isospin group. Isospin finds its most important applications in particle physics. But it was first discussed in nuclear physics.

Nuclear physics studies the way neutrons, N, and protons, P, bind together to form nucleii. There are three forces at work in the nucleus (ignoring gravity). Strong forces bind the nucleons (P or N) together. Electromagnetic (EM) interactions cause the P's to repel one another. Weak interactions cause β-decays of unstable nuclei.

A lot of physics can be explained by the simple assumption that the strong force between nucleons is independent of particle type - that it is the same for PP, PN or NN.

Suppose we could turn off the EM and weak interactions and at the same time eliminate the small mass difference between P and N, then look at any nucleus. If we take out any N and replace it with a P, all the forces are the same, so we ought to get a state with the same energy, right? Almost but not quite, because the N's are identical particles, so we have to sum over all possible replacements, $N \rightarrow P$.

To avoid getting lost in particle identities, it is convenient to introduce creation and annihilation operators. In non-relativistic quantum mechanics, we are used to thinking of the physical Hilbert space as describing a fixed number of particles. But we can put together spaces describing different numbers of particles into a single

Hilbert space and consider operators which take us from one space to another.

Let $P^\dagger_\alpha (P_\alpha)$ be an operator which creates (annihilates) a P in a state α (where α runs over all necessary labels, say spin and position). $P^\dagger_\alpha = (P_\alpha)^\dagger$ since if

$$P^\dagger_\alpha |s> = |s + P \text{ in state } \alpha>$$

$$P_\alpha |s + P \text{ in state } \alpha> = |s>.$$

Because P's are fermions, they obey the Pauli exclusion principle, so we must have $P^\dagger_\alpha P^\dagger_\alpha = P_\alpha P_\alpha = 0$. Furthermore, $P^\dagger_\alpha P_\alpha + P_\alpha P^\dagger_\alpha = 1$, because the first term is 1 and the second zero acting on a state with a P in state α, while the first is zero and second 1 on a state with no P in state α. In general, we can choose

$$\{P^\dagger_\alpha, P_\beta\} = \delta_{\alpha\beta}$$
$$\{P^\dagger_\alpha, P^\dagger_\beta\} = \{P_\alpha, P_\beta\} = 0 \tag{V.1}$$

where $\{A, B\} = AB + BA$.

Call the state with no particles $|0>$. It satisfies $P_\alpha |0> = 0$ for all α. An n proton state is

$$P^\dagger_{\alpha_1} \cdots P^\dagger_{\alpha_n} |0> = |\alpha_1, \cdots, \alpha_n> \tag{V.2}$$

It satisfies the <u>Fermi-Dirac statistics</u>; that is it is antisymmetric in the exchange of labels of the identical particles, $\alpha_i \leftrightarrow \alpha_j$. This is a great convenience in dealing with states involving identical particles.

Similarly we can define $N^\dagger_\alpha (N_\alpha)$ which create (annihilate) N's in a state α. They can be chosen to satisfy

$$\{N^\dagger_\alpha, N_\beta\} = \delta_{\alpha\beta}$$
$$\{N^\dagger_\alpha, N^\dagger_\beta\} = \{N_\alpha, N_\beta\} = 0 \tag{V.3}$$

$$\{P^\dagger_\alpha \text{ (or } P_\alpha), N^\dagger_\beta \text{ (or } N_\beta)\} = 0. \qquad \text{(V.3)}$$

The creation and annihilation operators allow us to treat states with different numbers of particles on the same footing. We can build the Hamiltonian out of creation and annihilation operators. For example, the electromagnetic repulsion between P's might look like

$$\sum_{\alpha,\beta} P^\dagger_\alpha P_\alpha V_{\alpha\beta} P^\dagger_\beta P_\beta. \qquad \text{(V.4)}$$

Here $P^\dagger_\alpha P_\alpha$ counts the number of protons in the state α. It is equal to one when acting on a state vector with one proton in the state α, because the P_α annihilates the proton and the $P^\dagger_\alpha$ regenerates it giving back the same state. But it vanishes on a state vector with no protons in the state α because P_α has nothing to annihilate. The $V_{\alpha\beta}$ in (V.4) would thus be the electromagnetic interaction between a proton in state α and another in state β.

Now consider the operator

$$\sum_\alpha P^\dagger_\alpha N_\alpha. \qquad \text{(V.5)}$$

Acting on a state

$$|\alpha_1 \cdots \alpha_n; \beta_1 \cdots \beta_m>$$
$$= P^\dagger_{\alpha_1} \cdots P^\dagger_{\alpha_n} N^\dagger_{\beta_1} \cdots N^\dagger_{\beta_m} |0> \qquad \text{(V.6)}$$

this operator contributes only if $\alpha = \beta_i$, in which case an N in the state β_i is replaced by a proton in the same state. This is the operator we want which replaces a neutron by a proton. In the limit in which weak and EM interaction and the P-N mass difference are turned off, this operator should not change the energy. In other words, if we write the Hamiltonian

$$H = H_S + H_{EM} + H_W \tag{V.7}$$

where H_{EM} includes the effect of P-N mass difference and electromagnetic interaction and H_W describes the weak interactions while H_S describes the strong force, then

$$[H_S\, , \sum_\alpha P^\dagger_\alpha N_\alpha] = 0. \tag{V.8}$$

Similarly the adjoint operator $\sum_\alpha N^\dagger_\alpha P_\alpha$, which replaces a proton by a neutron, commutes with H_S.

Define

$$T^+ \equiv \frac{1}{\sqrt{2}} \sum_\alpha P^\dagger_\alpha N_\alpha \tag{V.9}$$

$$T^- \equiv \frac{1}{\sqrt{2}} \sum_\alpha N^\dagger_\alpha P_\alpha .$$

Now the commutator is

$$[T^+, T^-] \equiv T_3 \tag{V.10}$$

$$= \frac{1}{2} [\sum_\alpha P^\dagger_\alpha N_\alpha, \sum_\beta N^\dagger_\beta P_\beta]$$

$$= \frac{1}{2} \sum_{\alpha,\beta} (P^\dagger_\alpha N_\alpha N^\dagger_\beta P_\beta - N^\dagger_\beta P_\beta P^\dagger_\alpha N_\alpha)$$

$$+ P^\dagger_\alpha N^\dagger_\beta N_\alpha P_\beta - N^\dagger_\beta P^\dagger_\alpha P_\beta N_\alpha$$

$$= \frac{1}{2} \sum_{\alpha,\beta} (P^\dagger_\alpha P_\beta \delta_{\alpha\beta} - N^\dagger_\beta N_\alpha \delta_{\alpha\beta})$$

$$= \frac{1}{2} \sum_\alpha (P^\dagger_\alpha P_\alpha - N^\dagger_\alpha N_\alpha).$$

The operator T_3 counts the number of protons in a state, subtracts the neutron number and divides by two. Physically it is obvious that it commutes with H_S (and H_{EM}) because the interaction in a nucleus doesn't change the number of protons or neutrons. But more formally we can use the (V.10), (V.8) and its adjoint, and the Jacobi identity to compute the commutator,

$$[H_S, T_3] = [H_S, [T^+, T^-]] = 0. \qquad (V.11)$$

Using the same manipulations as in (V.10), we find

$$[T_3, T^{\pm}] = \pm T^{\pm}. \qquad (V.12)$$

These operators satisfy the angular momentum algebra! They are called the isospin generators. We can take over all our results from the angular momentum algebra and apply them unchanged to isospin.

We can, if we feel like it, define hermitian isospin generators T_3 and

$$T_1 = \frac{1}{\sqrt{2}}(T^+ + T^-) \quad \text{and} \quad T_2 = \frac{1}{i\sqrt{2}}(T^+ - T^-),$$

and we can exponentiate them to generate a continuous SU(2) symmetry group, just as we exponentiate the angular momentum operators to generate rotations. But here the group elements are quite useless. They make unitary transformations on the two dimensional space (P, N). But no one is interested in states which are linear combinations of P's and N's. States like

$$\frac{1}{\sqrt{2}}(P^{\dagger}_{\alpha} + N^{\dagger}_{\alpha})|0\rangle$$

do not correspond to physical states because they violate superselection rules. The physically relevant objects are the generators themselves, $T^{\pm}$ and T_3.

What can we do with isospin? Let's analyze the eigenstates of $H = H_S + H_{EM} + H_W$ in perturbation theory, taking H_S as the zeroth-order Hamiltonian and treating H_{EM} and H_W as small perturbations. The perturbation theory should converge rapidly because H_{EM} and H_W are really small. H_{EM} is characterized by the dimensionless number

$\alpha = 1/137$ ($\hbar = c = 1$) and H_W by the dimensional constant $G_F \simeq 10^{-5}/m_p^2$. Aside on units: We will always take $\hbar = c = 1$ and measure all dimensional quantities in MeV (million electron volts). $m_p \simeq 938$ MeV. In practice we will never go beyond first order in α and G_F.

Consider the subspace of eigenstates of H_S with eigenvalue h. Operating in any state in this subspace with an isospin generator T_3, $T^{\pm}$ gives another state in the subspace, because $[T_a, H_S] = 0$. So the subspace transforms according to some representation of the isospin algebra. If necessary, we can decompose it into its irreducible components. So the states of the system can be labeled according to their isospin properties.

$$|I_3, I, h, \alpha\rangle$$

where I_3 is the eigenvalue of T_3, I is the highest T_3 eigenvalue of the irreducible representation, h is the H_S eigenvalue and α is any other label necessary to specify the state. The $2I+1$ states with the same I, h and α are said to form an isospin multiplet with isospin I.

Now we can use perturbation theory to construct eigenstates of H, corresponding to the eigenstates $|I_3, I, h, \alpha\rangle$ of H_S. States in a given multiplet will not have exactly the same H eigenvalues (energies), but because the perturbation is small, the energy differences within multiplets (coming only from H_{EM} and H_W) will typically be much smaller than the splittings between multiplets (corresponding to different eigenvalues of H_S). So the first thing we learn is that the states fall into approximately degenerate isospin multiplets.

Examples: P, N - isospin 1/2. D(deuterium or H^2) - isospin 0. H^3 (tritium) and He^3 - isospin 1/2.

We say that isospin is conserved by the strong interactions because when weak and EM interactions are ignored, the isospin properties of a state do not change with time. This is clear because the time evolution of a state is governed by the operator $\exp\{-i H_S t\}$ which commutes with T_a. There are two other operators which are conserved by all known interactions: the charge Q and the baryon number B. In nuclear physics, the charge is just the number of P's

$$Q = \sum_\alpha P^\dagger_\alpha P_\alpha \tag{V.13}$$

and the baryon number is the number of nucleons

$$B = \sum_\alpha (P^\dagger_\alpha P_\alpha + N^\dagger_\alpha N_\alpha). \tag{V.14}$$

Clearly Q, B and T_3 are related by

$$Q = T_3 + \frac{1}{2} B. \tag{V.15}$$

In nuclear physics, isospin is rather trivial. It's just an elegant way of stating the simple fact that nuclei are bound states of P's and N's, with the binding interaction approximately independent of which type of particle is involved. The isospin generators have a simple physical interpretation; they turn a $P \to N$ or $N \to P$. But when you examine the strong interactions at very short distances by slamming nucleons together with energies comparable to their masses, things get more interesting. This is the domain of particle physics.

The first thing that happens when you crunch nucleons is that you produce pions (π^+, π^0, π^-). Pions have a mass of about 140 MeV and spin zero and baryon number zero. Particle physicists call them stable particles because they live such a long time, $\sim 10^{-8}$ sec for π^+

and 10^{-16} sec for π^0. That is a long time really, because the time required to produce a pion by the strong interactions is ~ 10^{-24} sec, the time required for light to traverse a proton radius. The decays of the pions are caused by $H_{EM}(\pi^0)$ and $H_W(\pi^{\pm})$. The π's are not built out of P's and N's in any obvious way, but isospin is still good.

The pions have $I = 1$, $\pi^{\pm}$ have $I_3 = \pm 1$ and π^0 has $I_3 = 0$. Define creation and annihilation operators $a_{\pi^+}^{\dagger}$, $a_{\pi^0}^{\dagger}$ and $a_{\pi^-}^{\dagger}$ (a_{π^+} etc.) which are just like $P^{\dagger}$ and $N^{\dagger}$ except that they obey commutation rather than anticommutation relations (and I have suppressed the α label). The isospin generators contain an additional term for pions (with a sum over the suppressed α label understood.)

$$
\begin{aligned}
T_3 &= a_{\pi^+}^{\dagger} a_{\pi^+} - a_{\pi^-}^{\dagger} a_{\pi^-} \\
T^+ &= a_{\pi^+}^{\dagger} a_{\pi^0} + a_{\pi^0}^{\dagger} a_{\pi^-} \qquad \text{(V.16)} \\
T^- &= a_{\pi^-}^{\dagger} a_{\pi^0} + a_{\pi^0}^{\dagger} a_{\pi^+} \\
&\quad + \text{P, N contributions.}
\end{aligned}
$$

SCATTERING

Even though $\pi^{\pm}$ last only ~ 10^{-8} sec before decaying, experimental physicists can make beams of them (relativity helps!) and shoot them at nucleons at rest. Sometimes they collide and scatter. In a scattering experiment, weak and EM interactions can be ignored, because the particles are only in "contact" for a very short time (~ 10^{-24} sec). So the isospin properties of the final state, after collision, are the same as those of the initial state. To study scattering in detail, it's useful to define the S-matrix. In a scattering experiment, the physicist prepares an initial state with a pion and proton far apart and not interacting, $|\pi\, P, I\rangle$. This state then evolves with time according to the laws of quantum mechanics. If the state is prepared at time t_I, at a later time t, the state is

$$e^{-iH(t-t_I)}|\pi\, P,\, I\rangle.$$

At some time t_F, long after the collision (if any) has occurred, the state will be a linear combination of states describing pions and nucleons coming out at various angles and energies. The coefficient of any particular final state $|\pi\, P,\, F\rangle$ is the S-matrix element, or scattering amplitude

$$S_{FI} = \langle\pi\, P,\, F|e^{-iH(t_F-t_I)}|\pi\, P,\, I\rangle. \tag{V.17}$$

The probability of producing the final state is $|S_{FI}|^2$. Ignoring weak and EM interaction, the time evolution operator is $\exp\{-i\, H_S t\}$, which commutes with the isospin generators.

Consider the following observable scattering processes: $\pi^+ P \to \pi^+ P$, $\pi^+ N \to \pi^+ N$, $\pi^+ N \to \pi^0 P$, $\pi^- P \to \pi^- P$, and $\pi^- N \to \pi^- N$ (π^0 decay in 10^{-16} sec, so even clever experimentalists cannot make beams of them, thus only $\pi^{\pm}$ are shown in initial states--and the $\pi^0 N$ final state is ignored because the two neutral particles are hard to see). These five processes are described by only two independent scattering amplitudes, one for isospin 3/2, and one for isospin 1/2. The reason is that the pion-nucleon states, as far as the isospin algebra is concerned, transform like the direct product of the isospin one pion representation and the isospin 1/2 nucleon representation. The total isospin is therefore 3/2 and 1/2, which is to say that any pion nucleon state can be written as a linear combination of an isospin 3/2 state and an isospin 1/2 state. Because isospin is conserved, the 3/2 and 1/2 states scatter independently.

Suppose, for example, that in π^+N scattering, the scattering amplitude to produce π^+N at a particular set of angles and energies is G,

$$\langle\pi^+N,\ F|S|\pi^+N,\ I\rangle = G \tag{V.18}$$

where $S = \exp\{i\ H_S(t_F - t_I)\}$. Further, suppose that the scattering amplitude into π^0P at the same angles and energies is B,

$$\langle\pi^0P,\ F|S|\pi^+N,\ I\rangle = B. \tag{V.19}$$

Let's calculate the scattering amplitude for $\pi^+P \rightarrow \pi^+P$ at the same angles and energies.

The state $|\pi^+P,\ I\rangle$ is pure isospin 3/2 with $T_3 = 3/2$,

$$|\pi^+P,\ I\rangle = |\tfrac{3}{2},\ \tfrac{3}{2},\ I\rangle. \tag{V.20}$$

The scattering amplitude we want is

$$\begin{aligned}&\langle\pi^+P,\ F|S|\pi^+P,\ I\rangle\\ &= \langle\tfrac{3}{2},\ \tfrac{3}{2},\ F|S|\tfrac{3}{2},\ \tfrac{3}{2},\ I\rangle \equiv a_{3/2}.\end{aligned} \tag{V.21}$$

But because $[S,\ T_a] = 0$, the scattering amplitudes must satisfy

$$\begin{aligned}&\langle m,\ j,\ F|S|m',\ j',\ I\rangle\\ &\qquad = \delta_{mm'}\ \delta_{jj'}\ a_j\end{aligned} \tag{V.22}$$

by the same arguments which lead to (III.12). Applying T^- to (V.20), we find

$$|\tfrac{1}{2},\ \tfrac{3}{2},\ I\rangle = \sqrt{\tfrac{2}{3}}\ |\pi^0P,\ I\rangle + \sqrt{\tfrac{1}{3}}\,|\pi^+N,\ I\rangle. \tag{V.23}$$

The orthogonal linear combination must be isospin 1/2,

$$|\tfrac{1}{2},\ \tfrac{1}{2},\ I\rangle = \sqrt{\tfrac{2}{3}}\,|\pi^+N,\ I\rangle - \sqrt{\tfrac{1}{3}}\ |\pi^0P,\ I\rangle. \tag{V.24}$$

In terms of isospin states (from (V.23 and 24))

$$|\pi^0 P\rangle = \sqrt{\frac{2}{3}}\,|\frac{1}{2}, \frac{3}{2}\rangle - \sqrt{\frac{1}{3}}\,|\frac{1}{2}, \frac{1}{2}\rangle$$

$$|\pi^+ N\rangle = \sqrt{\frac{2}{3}}\,|\frac{1}{2}, \frac{1}{2}\rangle + \sqrt{\frac{1}{3}}\,|\frac{1}{2}, \frac{3}{2}\rangle. \qquad (V.25)$$

So

$$G = \langle\pi^+ N, F|S|\pi^+ N, I\rangle = \frac{2}{3}\, a_{1/2} + \frac{1}{3}\, a_{3/2}$$

$$B = \langle\pi^0 P, F|S|\pi^+ N, I\rangle = \frac{\sqrt{2}}{3}\,(a_{3/2} - a_{1/2}). \qquad (V.26)$$

THE GENERALIZED EXCLUSION PRINCIPLE

It's convenient to think of the particles in a single isospin multiplet as identical particles, treating the T_3 value as just another label for the state, like T_3 or position. Indeed, this is just what we had in mind in choosing the creation and annihilation operators for P's to anticommute with those for N's. Write

$$a^\dagger_{1/2\alpha} = P^\dagger_\alpha, \quad a^\dagger_{-1/2\alpha} = N^\dagger_\alpha. \qquad (V.27)$$

Then the a's satisfy

$$\{a^\dagger_{i\alpha}, a_{j\beta}\} = \delta_{ij}\delta_{\alpha\beta}$$

$$\{a^\dagger_{i\alpha}, a^\dagger_{j\beta}\} = \{a_{i\alpha}, a_{j\beta}\} = 0. \qquad (V.28)$$

Consider a state like the deuteron of one P and one N. We can build such states from linear combinations of states with definite J_3, T_3 and position for each particle

$$|m, I_3, r; m', I_3', r'\rangle$$

$$= -|m', I_3', r'; m, I_3, r\rangle,$$

the - sign a consequence of the above anticommutation relations. Consider the total spin and isospin of such a state. Each is 1 or 0 obviously. But the J (or I) = 1 state is symmetric with respect to exchange of the spin (or isospin) labels of the two particles while the J (or I) = 0 state is antisymmetric. There is a relation between the total spin and isospin and the symmetry of the state with respect to interchange of particle positions. If the spatial wave function is symmetric, only $J = 1$, $I = 0$ or $J = 0$, $I = 1$ is possible, while if the spatial wave function is antisymmetric, we must have $J = I$. This is an example of the generalized Pauli exclusion principle.

A MORE ELEGANT NOTATION

Notice that in this new notation the isospin generators become

$$T_3 = \frac{1}{2} \sum_\alpha (a^\dagger_{1/2\alpha}\, a_{1/2\alpha} - a^\dagger_{-1/2\alpha}\, a_{-1/2\alpha})$$

$$= \frac{1}{2}\, a^\dagger_{i\alpha}\, [\sigma_3]_{ij}\, a_{j\alpha}. \tag{V.29}$$

In fact, you can check that

$$T_a = \sum_{\substack{\alpha \\ i,j}} a^\dagger_{i\alpha}\, [\sigma_a]_{ij}\, a_{j\alpha}. \tag{V.30}$$

This simple form is quite general. Suppose the creation (annihilation) operators for an isomultiplet of particles with isospin I are $b^+_{i\alpha}$ $(b_{i\alpha})$ for $i = -I$ to I. They satisfy

$$[b_{i\alpha}, b^\dagger_{j\beta}]_\pm$$

$$\equiv b_{i\alpha} b^\dagger_{j\beta} \pm b^\dagger_{j\beta} b_{i\alpha} = \delta_{ij}\, \delta_{\alpha\beta}$$

$$[b^\dagger_{i\alpha}, b^\dagger_{j\beta}]_\pm = [b_{i\alpha}, b_{j\beta}]_\pm = 0. \tag{V.31}$$

Then the isospin generators are

$$T_a = \sum_\alpha b^\dagger_{i\alpha} [T^I_a]_{ij}\, b_{j\alpha}. \tag{V.32}$$

You will show in the exercizes that these obey the appropriate commutation relations.

PROBLEMS FOR CHAPTER V

(V.A) Suppose that in some process, a pair of pions are produced in an $\ell=0$ state. What total isospin values are possible for this state?

(V.B) Suppose X_a are NxN matrices satisfying

$$[X_a, X_b] = i\, f_{abc} X_c$$

and $b_i^\dagger(b_i)$ are creation and annihilation operators satisfying (i=1 to N)

$$[b_i, b_j^\dagger]_\pm = \delta_{ij}$$

$$[b_i^\dagger, b_j^\dagger]_\pm = [b_i, b_j]_\pm = 0.$$

Show that the operators

$$\chi_a \equiv \sum_{i,j} b_i^\dagger [X_a]_{ij}\, b_j$$

satisfy

$$[\chi_a, \chi_b] = i\, f_{abc} \chi_c.$$

(V.C) Δ^{++}, Δ^{+}, Δ^{0} and Δ^{-} are an isospin 3/2 multiplet of particles (T_3 = 3/2, 1/2, -1/2, -3/2 respectively) with baryon number one. They are produced by strong interactions in π-nucleon collisions. Compare the probability of producing Δ^{++} in $\pi^+ P \to \Delta^{++}$ with the probability of producing Δ^0 in $\pi^- P \to \Delta^0$.

VI. ROOTS AND WEIGHTS

In this chapter, we leave physics for a while. We go back to mathematics to generalize the discussion of SU(2) to an arbitrary simple Lie algebra, following Cartan and Dynkin.

The idea is to divide the generators into two sets: one set the analog J_3 in the angular momentum algebra, hermitian operators to be diagonalized; the other set the analog of $J^{\pm}$ in angular momentum, raising and lowering operators.

Howard Georgi, Lie Algebras in Particle Physics: From Isospin to Unified Theories ISBN 0-8053-3153-0

Suppose X_a are the generators of a simple Lie algebra, in an irreducible representation D. We will analyze it in analogy with SU(2). The first step is to diagonalize as many generators as we can. We choose linear combinations of the X_a, H_i for i=1 to m, with the following properties:

H_i is Hermitian $= C_{ia}X_a$, C real;

$[H_i, H_j] = 0;$

$$tr(H_iH_j) = k_D\delta_{ij}; \tag{VI.1}$$

m is as large as possible.

These generators are the analog of J_3, a maximal set of commuting Hermitian operators whose eigenvalues can be used to label the states of the representation D. The integer m is called the <u>rank</u> of the algebra, and of the group it generates.

Now we diagonalize H_i and write the states of the representation as $|\mu, D>$ where

$$H_i|\mu, D> = \mu_i|\mu, D>. \tag{VI.2}$$

The set H_i is called the <u>Cartan subalgebra</u>. The eigenvalues μ_i are called <u>weights</u> and the vector μ with components μ_i is a <u>weight vector</u>.

ROOTS

We define the weights for any representation using

the same generators H_i. In particular, we can analyze the adjoint representation. In the adjoint representation, (II.6), the states correspond to generators $X_a \to |X_a\rangle$ with the scalar product (see II.8)

$$\langle X_a | X_b \rangle = \lambda^{-1} \operatorname{tr}(X_a^\dagger X_b). \tag{VI.3}$$

The action of the generators on these states is

$$X_a | X_b \rangle = | [X_a, X_b] \rangle \tag{VI.4}$$

We already know about a set of states with weight zero, namely $|H_i\rangle$

$$H_i | H_j \rangle = 0. \tag{VI.5}$$

Diagonalizing the rest of the space gives states $|E_\alpha\rangle$ such that

$$H_i | E_\alpha \rangle = \alpha_i | E_\alpha \rangle \tag{VI.6}$$

corresponding to generators E_α,

$$[H_i, E_\alpha] = \alpha_i E_\alpha. \tag{VI.7}$$

These generators are necessarily not hermitian

$$[H_i, E_\alpha^\dagger] = -\alpha_i E_\alpha^\dagger, \tag{VI.8}$$

so $E_\alpha^\dagger = E_{-\alpha}$. We will normalize

$$\langle E_\alpha | E_\beta \rangle = \lambda^{-1} \operatorname{tr}(E_\alpha^\dagger E_\beta) = \delta_{\alpha\beta} \tag{VI.9}$$

$$\langle H_i | H_j \rangle = \lambda^{-1} \operatorname{tr}(H_i H_j) = \delta_{ij}.$$

Comparing (VI.9) and (VI.1), we see that $\lambda = k_D$ when D is the adjoint representation. It is standard in the mathematics literature to take $\lambda = 1$. In physics, this is often inconvenient, so we will ignore the convention and choose λ in each group for convenience. Once λ is fixed, all the other k_Ds are determined.

The weights, α_i, of the adjoint representation are called roots.

RAISING AND LOWERING OPERATORS

The E_α are raising and lowering operators for the weights (like $J^\pm$ in SU(2)).

$$H_i E_\alpha |\mu, D\rangle = [H_i, E_\alpha] |\mu, D\rangle$$
$$+ E_\alpha H_i |\mu, D\rangle = (\mu+\alpha)_i E_\alpha |\mu, D\rangle. \qquad \text{(VI.10)}$$

Starting with any state $|\mu, D\rangle$, we can derive properties of the roots and weights which are analogous to the statement that for SU(2) the eigenvalues of J_3 are half-integers. If there is more than one state with weight μ, we will choose $|\mu, D\rangle$ to be an eigenstate of $E_{-\alpha}E_\alpha$ for some fixed α, and define

$$E_{\pm\alpha}|\mu, D\rangle = N_{\pm\alpha,\mu}|\mu\pm\alpha, D\rangle. \qquad \text{(VI.11)}$$

Now, in the adjoint representation the state $E_\alpha|E_{-\alpha}\rangle$ has weight zero, so

$$E_\alpha|E_{-\alpha}\rangle = \beta_i|H_i\rangle,$$
$$\beta_i = \langle H_i|E_\alpha|E_{-\alpha}\rangle$$
$$= \mathrm{tr}(H_i[E_\alpha, E_{-\alpha}])/\lambda$$
$$= \mathrm{tr}(E_{-\alpha}[H_i, E_\alpha])/\lambda = \alpha_i. \qquad \text{(VI.12)}$$

Thus

$$[E_\alpha, E_{-\alpha}] = \alpha_i H_i. \qquad \text{(VI.13)}$$

Now back to the representation D. Consider

$$\langle\mu, D|[E_\alpha, E_{-\alpha}]|\mu, D\rangle$$
$$= \langle\mu, D|\alpha_i H_i|\mu, D\rangle = \alpha\cdot\mu = \langle\mu, D|E_\alpha E_{-\alpha}|\mu, D\rangle$$
$$- \langle\mu, D|E_{-\alpha}E_\alpha|\mu, D\rangle = |N_{-\alpha,\mu}|^2 - |N_{\alpha,\mu}|^2. \qquad \text{(VI.14)}$$

But

$$N_{-\alpha,\mu} = \langle\mu-\alpha,\ D|E_{-\alpha}|\mu,\ D\rangle$$
$$= \langle\mu-\alpha,\ D|E_{\alpha}^{\dagger}|\mu,\ D\rangle$$
$$= \langle\mu,\ D|E_{\alpha}|\mu-\alpha,\ D\rangle^{*} = N_{\alpha,\mu-\alpha}^{*}. \qquad \text{(VI.15)}$$

So

$$|N_{\alpha,\mu-\alpha}|^2 - |N_{\alpha,\mu}|^2 = \alpha\cdot\mu. \qquad \text{(VI.16)}$$

If we apply E_α or $E_{-\alpha}$ repeatedly to $|\mu,\ D\rangle$, we must eventually get zero. Suppose

$$E_\alpha|\mu+p\alpha,\ D\rangle = 0, \quad E_{-\alpha}|\mu-q\alpha,\ D\rangle = 0, \qquad \text{(VI.17)}$$

for positive integers p and q. Then

$$|N_{\alpha,\mu+(p-1)\alpha}|^2 - 0 = \alpha\cdot(\mu+p\alpha)$$
$$|N_{\alpha,\mu+(p-2)\alpha}|^2 - |N_{\alpha,\mu+(p-1)\alpha}|^2 = \alpha\cdot(\mu+(p-1)\alpha)$$
$$\vdots$$
$$|N_{\alpha,\mu}|^2 - |N_{\alpha,\mu+\alpha}|^2 = \alpha\cdot(\mu+\alpha)$$
$$|N_{\alpha,\mu-\alpha}|^2 - |N_{\alpha,\mu}|^2 = \alpha\cdot\mu$$
$$\vdots$$
$$|N_{\alpha,\mu-q\alpha}|^2 - |N_{\alpha,\mu-(q-1)\alpha}|^2 = \alpha\cdot(\mu-(q-1)\alpha)$$
$$0 - |N_{\alpha,\mu-q\alpha}|^2 = \alpha\cdot(\mu-q\alpha)$$

$$0 = (p+q+1)(\alpha\cdot\mu) + \alpha^2\left[\frac{p(p+1)}{2} - \frac{q(q+1)}{2}\right]$$
$$= (p+q+1)\left\{(\alpha\cdot\mu) + \frac{\alpha^2}{2}(p-q)\right\}. \qquad \text{(VI.18)}$$

So the combination $\alpha\cdot\mu/\alpha^2$ is a half integer:

$$\frac{\alpha\cdot\mu}{\alpha^2} = -\frac{1}{2}(p-q). \qquad \text{(VI.19)}$$

Note, also, that we can determine all the $|N|^2$ in terms of μ, α, p and q.

This is true for any representation, but it has particularly simple and strong consequences for the adjoint representation. There μ is a root, say β, and we have

$$\frac{\alpha\cdot\beta}{\alpha^2} = -\frac{1}{2}(p-q) = \frac{m}{2} \tag{VI.20}$$

for some integer m. But we can equally well apply $E_{\pm\beta}$ to the state $|E_\alpha>$ and obtain

$$\frac{\beta\cdot\alpha}{\beta^2} = -\frac{1}{2}(p'-q') = \frac{m'}{2}. \tag{VI.21}$$

Multiplying (VI.20 and 21) we find

$$\frac{mm'}{4} = \frac{(\alpha\cdot\beta)^2}{\alpha^2\beta^2} = \cos^2\theta \tag{VI.22}$$

where θ is the angle between the root vectors. The only possibilities are

mm'	θ
0	90°
1	60°, 120°
2	45°, 135°
3	30°, 150°
4	0°, 180°

(VI.23)

Only these angles are allowed between roots of simple Lie algebras.

We've been assuming in our notation that for each non-zero root vector, α, there is a unique generator E_α. We can now prove it easily. Suppose E_α and E'_α are different generators with the same root. We can choose them to be orthogonal

$$<E_\alpha|E'_\alpha> = \lambda^{-1}\, \mathrm{tr}(E_\alpha^\dagger E'_\alpha) = 0. \tag{VI.24}$$

Applying $E_{\pm\alpha}$ to the state $|E'_\alpha\rangle$, we get the relation

$$\frac{\alpha\cdot\alpha}{\alpha^2} = -\frac{1}{2}(p-q). \tag{VI.25}$$

But $q = 0$ because

$$E_{-\alpha}|E'_\alpha\rangle = \beta_i H_i$$

$$\beta_i = \langle H_i|E_{-\alpha}|E'_\alpha\rangle = \mathrm{tr}(H_i[E_{-\alpha}, E'_\alpha])/\lambda$$

$$= \mathrm{tr}(E'_\alpha[H_i, E_{-\alpha}])/\lambda = -\alpha_i\ \mathrm{tr}(E'_\alpha E_{-\alpha})/\lambda = 0. \tag{VI.26}$$

So

$$\frac{\alpha\cdot\alpha}{\alpha^2} = -\frac{1}{2}p \tag{VI.27}$$

which is impossible because p is positive. E_α is unique.

The analysis which leads to (VI.19) is essentially the same one we used in Chapter III to determine the representations of SU(2). The connection with SU(2) may be more clear in the following alternative derivation.

Define the rescaled generators

$$E^\pm \equiv |\alpha|^{-1} E_{\pm\alpha} \quad (|\alpha|^2 = \alpha\cdot\alpha),$$

$$E_3 = |\alpha|^{-2} \alpha_i H_i. \tag{VI.28}$$

You can use (VI.7, 8 and 13) to show that they satisfy the SU(2) algebra

$$[E_3, E^\pm] = \pm E^\pm; \quad [E^+, E^-] = E_3. \tag{VI.29}$$

Thus, each E_α is associated with an SU(2) subalgebra comprising (with the rescalings in (VI.28)) E_α, its Hermitian adjoint $E_{-\alpha}$ and a linear combination of Cartan generators.

If $|\mu, D\rangle$ is an eigenstate of $E_{-\alpha}E_\alpha$, then the states

$$|\mu+k\alpha, D\rangle \tag{VI.30}$$

obtained by applying $E_{\pm\alpha}$ to $|\mu, D\rangle$ form an irreducible representation of the SU(2) subalgebra (VI.29). The condition that $|\mu, D>$ is an eigenstate of $E_{-\alpha}E_{\alpha}$ is necessary to insure that the state is not a linear combination of states with the same μ from different representations. These states satisfy

$$E_3|\mu+k\alpha, D> = |\alpha|^{-2}(\alpha\cdot\mu+k\alpha^2)|\mu+k\alpha, D> \qquad \text{(VI.31)}$$

The highest E_3 value in the representation, corresponding to $|\mu+p\alpha, D>$, must be

$$j = \alpha\cdot\mu/\alpha^2+p \qquad \text{(VI.32)}$$

for the spin j representation. The lowest E_3 value, corresponding to $|\mu-q\alpha, D>$, is then

$$-j = \alpha\cdot\mu/\alpha^2-q. \qquad \text{(VI.33)}$$

Adding (VI.32) and (VI.33) gives (VI.19).

PROBLEMS FOR CHAPTER VI

(VI.A) Prove (VI.4), using (II.6).

(VI.B) Show that $[E_\alpha, E_\beta]$ must be proportional to $E_{\alpha+\beta}$. What if $E_{\alpha+\beta}$ is not a root?

(VI.C) Consider the simple Lie algebra formed by the 10 matrices: σ_a, $\sigma_a\tau_1$, $\sigma_a\tau_3$ and τ_2 where σ_a and τ_a are Pauli matrices in orthogonal spaces (see problem (III.E)). Take $H_1 = \sigma_3$ and $H_2 = \sigma_3\tau_3$ as the Cartan subalgebra. Find

(a) the roots of the adjoint representation and

(b) the weights of this representation.

VII. SU(3)

Let us take a break from highbrow mathematics and illustrate the concepts of the previous chapter in a simple example. The example is the SU(3) algebra. Later, we will see that SU(3) has many uses in particle physics.

SU(3) is the group of 3 x 3 unitary matrices with determinant 1. The generators are the 3 x 3 hermitian traceless matrices. The tracelessness is required by the condition that the determinant is one. The standard basis (in the physics literature) consists of the Gell-Mann λ matrices:

Howard Georgi, Lie Algebras in Particle Physics: From Isospin to Unified Theories ISBN 0-8053-3153-0

$$\lambda_1 = \begin{vmatrix} 0 & 1 & 0 \\ 1 & 0 & 0 \\ 0 & 0 & 0 \end{vmatrix} \qquad \lambda_2 = \begin{vmatrix} 0 & -i & 0 \\ i & 0 & 0 \\ 0 & 0 & 0 \end{vmatrix}$$

$$\lambda_3 = \begin{vmatrix} 1 & 0 & 0 \\ 0 & -1 & 0 \\ 0 & 0 & 0 \end{vmatrix} \qquad \lambda_4 = \begin{vmatrix} 0 & 0 & 1 \\ 0 & 0 & 0 \\ 1 & 0 & 0 \end{vmatrix}$$

$$\lambda_5 = \begin{vmatrix} 0 & 0 & -i \\ 0 & 0 & 0 \\ i & 0 & 0 \end{vmatrix} \qquad \lambda_6 = \begin{vmatrix} 0 & 0 & 0 \\ 0 & 0 & 1 \\ 0 & 1 & 0 \end{vmatrix} \qquad \text{(VII.1)}$$

$$\lambda_7 = \begin{vmatrix} 0 & 0 & 0 \\ 0 & 0 & -i \\ 0 & i & 0 \end{vmatrix} \qquad \lambda_8 = \frac{1}{\sqrt{3}} \begin{vmatrix} 1 & 0 & 0 \\ 0 & 1 & 0 \\ 0 & 0 & -2 \end{vmatrix} .$$

The generators are $T_a = \lambda_a/2$. Notice that

$$\text{tr}(T_a T_b) = \frac{1}{2} \delta_{ab} . \qquad \text{(VII.2)}$$

T_1, T_2 and T_3 generate an SU(2) subgroup of SU(3) which is called the <u>isospin</u> subgroup, because in the physical application of SU(3) as an internal symmetry, it is isospin. It is convenient to choose T_3 as an element of the Cartan subalgebra. There is only one generator which commutes with T_3, that is T_8. So we take

$$H_1 = T_3, \quad H_2 = T_8. \tag{VII.3}$$

SU(3) is rank 2.

A good reason for choosing T_3 and T_8 as our Cartan-subalgebra is that Gell-Mann has already diagonalized them for us. The eigenvectors are

$$\begin{vmatrix} 1 \\ 0 \\ 0 \end{vmatrix} \rightarrow \text{the state } \left|\frac{1}{2}, \frac{1}{2\sqrt{3}}\right\rangle,$$

$$\begin{vmatrix} 0 \\ 1 \\ 0 \end{vmatrix} \rightarrow \left|-\frac{1}{2}, \frac{1}{2\sqrt{3}}\right\rangle, \tag{VII.4}$$

$$\begin{vmatrix} 0 \\ 0 \\ 1 \end{vmatrix} \rightarrow \left|0, -\frac{1}{\sqrt{3}}\right\rangle,$$

where the states are labeled by their weights, $|\mu_1, \mu_2\rangle$. We can plot the weight-vectors in the plane.

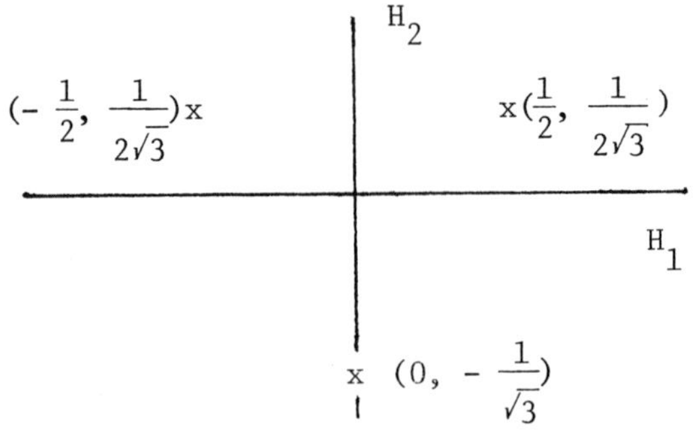

H_2

$(-\frac{1}{2}, \frac{1}{2\sqrt{3}})$ x   x $(\frac{1}{2}, \frac{1}{2\sqrt{3}})$

H_1 (VII.i)

x $(0, -\frac{1}{\sqrt{3}})$

They form an equilateral triangle!

We know the root vectors E_α are going to take us from one weight to another, so we can be fairly certain that the roots will be differences of weights (1, 0), $(1/2, \sqrt{3}/2)$, $(-1/2, \sqrt{3}/2)$ and minus these. To find the corresponding generators, we look for matrices that take us from one state to another. These are matrices with only a single off-diagonal element. Thus,

$$\frac{1}{\sqrt{2}}(T_1 \pm iT_2) = E_{\pm 1,0}$$

$$\frac{1}{\sqrt{2}}(T_4 \pm iT_5) = E_{1/2,\, \pm\sqrt{3}/2} \qquad \text{(VII.5)}$$

$$\frac{1}{\sqrt{2}}(T_6 \pm iT_7) = E_{\mp 1/2,\, \pm\sqrt{3}/2}$$

The roots form a regular hexagon

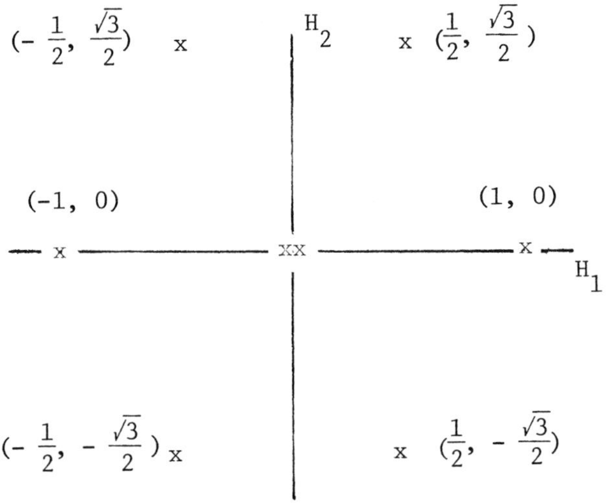

$(-\frac{1}{2}, \frac{\sqrt{3}}{2})$ x H_2 x $(\frac{1}{2}, \frac{\sqrt{3}}{2})$

(-1, 0) (1, 0) (VII.ii)

x xx x H_1

$(-\frac{1}{2}, -\frac{\sqrt{3}}{2})$ x x $(\frac{1}{2}, -\frac{\sqrt{3}}{2})$

All angles in SU(3) are multiples of 60°.

PROBLEMS FOR CHAPTER VII

(VII.A) Calculate f_{147} and f_{458} in SU(3).

(VII.B) Show that λ_2, λ_5 and λ_7 generate an SU(2) subalgebra of SU(3). How does the representation generated by (VII.1 and 2) transform under this subalgebra? That is, decompose (if necessary) the three dimensional representation into representations which are irreducible under this subalgebra.

(VII.C) Show that T_1, T_2 and T_3 generate an SU(2) subalgebra of SU(3). How does the representation generated by (VII.1 and 2) transform under this subalgebra?

VIII. SIMPLE ROOTS

We are now going to define a concept of "highest weight" for an arbitrary Lie group. It is based on a definition of positivity for weight vectors. We fix a basis in the Cartan subalgebra, H_1, H_2 ... . This fixes the components of the weight vectors, μ_1, μ_2... . A weight vector is said to be _positive_ if its first non-zero component is positive. Every weight vector is either positive, negative (if the first non-zero component is negative) or

Howard Georgi, Lie Algebras in Particle Physics: From Isospin to Unified Theories ISBN 0-8053-3153-0

zero (if all components are zero). This is an arbitrary division of the space into two halves.

It seems rather odd that such a seemingly random definition of positivity can do us any good at all. In fact, as we will see, it is a great convenience. It will allow us to talk about raising and lowering operators, for example. Even so, you might worry that the arbitrary ordering we have introduced will prejudice our notation in some confusing way. Indeed, we will have to show that this does not happen. We will eventually be able to see that it does not matter what definition of positivity we use, so long as we have one.

Now that we have a notion of positivity, we can define an ordering. We say $x > y$ if $x - y > 0$. Now obviously, the highest weight in an irreducible representation is a weight greater than any of the others. It will turn out that the highest weight is unique in general for the same reason that it is unique in SU(2): starting with the highest weight, you can build the whole representation.

We can apply the concept of positivity to the roots, the weights of the adjoint representation. The roots are either positive or negative. The positive roots are like raising operators, and negative roots like lowering operators. Obviously, if we act on the state of highest weight in any representation with E_α for positive α, we must get zero.

A _simple root_ is a positive root which cannot be written as the sum of two positive roots. We now show that the simple roots determine the whole structure of the group.

If α, β are simple roots, then $\beta - \alpha$ is not a root. Proof: assume $\beta - \alpha > 0$ is a root, then $\beta = \alpha + (\beta - \alpha)$, the sum of two positive roots and therefore β is not simple. Likewise if $\beta - \alpha < 0$ is a root, α is not simple.

So $E_{-\alpha}|E_\beta> = 0$, so in our master formula, (VI.19)

$$\frac{\alpha\cdot\beta}{\alpha^2} = -\frac{1}{2}(p-q)$$

the integer of q is zero. Thus,

$$\frac{2\alpha\cdot\beta}{\alpha^2} = -p. \tag{VIII.1}$$

Knowing all the integers p is equivalent to knowing the angle between each pair of simple roots and the relative lengths, because

$$\frac{2\alpha\cdot\beta}{\alpha^2} = -p, \quad \frac{2\alpha\cdot\beta}{\beta^2} = -p' \tag{VIII.2}$$

implies

$$\cos\theta = -\frac{1}{2}\sqrt{pp'}, \quad \frac{\beta^2}{\alpha^2} = \frac{p}{p'}. \tag{VIII.3}$$

The angle θ between any pair of simple roots satisfies

$$\frac{\pi}{2} \le \theta < \pi. \tag{VIII.4}$$

It is a simple geometrical fact that any set of vectors satisfying this constraint is linearly independent. The next paragraph is a dull proof of this fact.

The simple roots are linearly independent. If they were not, we could write

$$\sum_\alpha x_\alpha \alpha = 0 \quad \text{for some constants } x_\alpha. \tag{VIII.5}$$

Divide the simple roots into two sets $\Gamma_\pm$, $\alpha \in \Gamma_+$ if $x_\alpha \ge 0$ and $\alpha \in \Gamma_-$ if $x_\alpha < 0$. Then

$$y \equiv \sum_{\alpha\in\Gamma_+} x_\alpha \alpha = \sum_{\alpha\in\Gamma_-} (-x_\alpha)\alpha \equiv z \tag{VIII.6}$$

where all the constants are nonnegative. Both y and z are

positive (non-zero) vectors because they are sums of positive vectors. But this equality is impossible, since

$$y^2 = y \cdot z \leq 0 \qquad \text{(VIII.7)}$$

because $\alpha \cdot \beta \leq 0$.

Any positive root, ϕ, can be written as a sum of simple roots, α, with nonnegative integer coefficients, k_α,

$$\phi = \sum_\alpha k_\alpha \alpha. \qquad \text{(VIII.8)}$$

This can be seen as follows: if ϕ is simple, (VIII.8) is true. If not, I can split it into two positive roots ϕ_1 and ϕ_2. If they are both simple (VIII.8) is true. If either ϕ_1 or ϕ_2 is not simple, I can split it again into positive roots. Obviously, we can continue the process until we obtain (VIII.8).

The number of simple roots is m, equal to the rank of the group. Obviously, the number cannot be greater than m because they are linearly independent m-vectors. Suppose it is less than m, so that the simple roots do not span the whole space. Then, I can choose a basis in which the first component of all the α vanish. But that means the first component of every root vanishes, by (VIII.8) which means

$$[H_1, E_\phi] = 0 \text{ for all roots } \phi, \qquad \text{(VIII.9)}$$

but

$$[H_1, H_i] = 0 \quad \text{for all } i. \qquad \text{(VIII.10)}$$

Thus, H_1 commutes with everything, and it is an invariant subalgebra all by itself, which is impossible since we are assuming that the group is simple.

We can determine all the roots in terms of the simple roots if we can figure out which of the vectors

$$\sum_{\substack{\alpha \\ \text{simple}}} k_\alpha \alpha \qquad \text{(VIII.11)}$$

are actually roots. We can do this inductively. Let $k = \sum_\alpha k_\alpha$. For $k = 1$, the sums are just the simple roots themselves. Now, we obtain $k = 2$ by acting on the $k = 1$ sector with the generators corresponding to the simple roots and we use our master formula to determine whether we get a new root or zero. For example, $\alpha + \beta$ is a root if $\alpha \cdot \beta < 0$ ($\neq 0$) because then

$$\frac{2\alpha \cdot \beta}{\alpha^2} = -(p-q) = -p \Rightarrow p > 0 \qquad \text{(VIII.12)}$$

where p is the highest integer such that $\beta + p\alpha$ is a root. In this case we knew $q = 0$ because α and β were simple. But in all cases, $k = n$ acted on by a simple root α, q can be determined by examining the roots with $k < n$. And once we know q, we can use the master formula

$$\frac{2\alpha \cdot \gamma}{\alpha^2} = -(p-q) \qquad \text{(VIII.13)}$$

to determine p. If p is not zero, then $\gamma + \alpha$ is a root. In this way we determine all the roots with $k = n+1$ which can be written as $\gamma + \alpha$ where γ has $k = n$ and α is simple. But that is all the roots with $k = n+1$, for suppose there is some root ρ which does not have the form $\gamma + \alpha$ for any α. Then, since $\rho - \alpha$ is not a root, $q = 0$ and

$$\frac{2\alpha \cdot \rho}{\alpha^2} = -p \leq 0 \quad \text{for all } \alpha. \qquad \text{(VIII.14)}$$

Thus, ρ is linearly independent of all the simple roots, which is a contradiction.

Notice that we have determined all the roots inductively in terms of the simple roots, using only the lengths and angles between the simple roots and the master formula. This shows that the whole procedure is independent of the basis in root space.

Let's illustrate this procedure for SU(3). The positive roots are

$$(1,\ 0),\ (\tfrac{1}{2},\ \tfrac{\sqrt{3}}{2}) \quad \text{and} \quad (\tfrac{1}{2},\ -\tfrac{\sqrt{3}}{2}). \qquad \text{(VIII.15)}$$

(1,0) is not simple, since it is the sum of the other two. The simple roots are

$$\alpha = (\tfrac{1}{2},\ \tfrac{\sqrt{3}}{2}),\ \beta = (\tfrac{1}{2},\ -\tfrac{\sqrt{3}}{2}). \qquad \text{(VIII.16)}$$

Thus,

$$\alpha^2 = \beta^2 = 1,\ \alpha\cdot\beta = -\frac{1}{2},\ \text{so}$$

$$\frac{2\alpha\cdot\beta}{\alpha^2} = \frac{2\alpha\cdot\beta}{\beta^2} = -1 = -(p-q). \qquad \text{(VIII.17)}$$

So consider $E_\alpha|E_\beta>$. Since $p = 1$, $\beta+\alpha$ is a root, but $\beta+2\alpha$ is not.

DYNKIN DIAGRAMS

Let's restate what we know. The m simple roots can be used to find all the roots, which in turn we can use to write down the algebra by finding the $N_{\alpha,\beta}$. The simple roots are all we need to know. A Dynkin diagram is just a shorthand, diagrammatic notation for writing down the simple roots. Each simple root is written as an open circle. Pairs of circles are then connected by lines depending on the angle between the pair of roots to which the circles correspond, as follows:

○≡○ if the angle is 150°

○=○ if the angle is 135°

○–○ if the angle is 120°

○ ○ if the angle is 90° (VIII.18)

The complete Dynkin Diagram determines all the angles between pairs of simple roots. This doesn't quite determine all the simple roots, because there may be several possible choices for the relative lengths, but we will worry about that later.

The diagram for SU(2) is ○, for SU(3) it ○–○.

FUNDAMENTAL WEIGHTS

Label the simple roots α^i, $i = 1$ to m. Now consider the highest weight of an arbitrary irreducible representation, D. A weight μ in D is the highest weight in the representation if and only if $\mu+\phi$ is not a weight in the representation for all positive roots ϕ. Clearly, it is sufficient to require $\mu+\alpha^i$ not a weight for all the simple roots. That means that for E_{α^i} acting on μ, $p=0$, thus

$$\frac{2\alpha^i \cdot \mu}{\alpha^{i2}} = q^i \qquad \text{(VIII.19)}$$

where the q^i's are non-negative integers.

Since the α^i's are linearly independent, the q^i's completely determine μ. Every set of q^i gives a vector μ satisfying (VIII.19) which is the highest weight of some irreducible representation, and we can construct the entire representation by acting on μ with the lowering operators $E_{-\alpha^i}$.

We have shown that the irreducible representations of a rank m simple Lie group can be labeled by a set of m nonnegative integers, q^i. To get a feeling for what this means, it is useful to consider the weight vectors μ^i, such that

$$\frac{2\alpha^i \cdot \mu^j}{\alpha^{i2}} = \delta^{ij} . \tag{VIII.20}$$

μ^i is the highest weight of a representation with $q^i = 1$ and $q^j = 0$ for $i \neq j$. Clearly, any highest weight μ, can be written uniquely as a sum of the μ^i's,

$$\mu = \sum_i q^i \mu^i . \tag{VIII.21}$$

We can build the representation with highest weight μ out of the tensor product of q^1 representations with highest weight μ^1, q^2 with highest weight μ^2, etc; just as we build the spin n/2 representation of SU(2) out of n spin 1/2 representations.

The vectors μ^i are called the <u>fundamental weights</u>, and the m irreducible representations whose highest weights are μ^i are the <u>fundamental representations</u>, D^i.

Do not confuse the upper indices which label the simple roots and the fundamental weights with the lower indices which label the components of all the weight and root vectors. Both indices run from 1 to the rank of the group, but the first is just a label while the second is a vector index.

PROBLEMS FOR CHAPTER VIII

(VIII.A) Find the simple roots and fundamental weights and the Dynkin diagram for the algebra discussed in problem (VI.C).

(VIII.B) Consider the algebra generated by σ_a and $\sigma_a \eta_1$ where σ_a and η_a are independent Pauli matrices. Show that this algebra generates a group which is semisimple but not simple. Nevertheless, you can define simple roots. What does the Dynkin diagram look like?

IX. MORE SU(3)

In SU(3), we can label the simple roots as

$$\alpha^1 = \left(\frac{1}{2}, \frac{\sqrt{3}}{2}\right), \quad \alpha^2 = \left(\frac{1}{2}, -\frac{\sqrt{3}}{2}\right). \tag{IX.1}$$

Then the μ^i are

$$\mu^1 = \left(\frac{1}{2}, \frac{1}{2\sqrt{3}}\right), \quad \mu^2 = \left(\frac{1}{2}, -\frac{1}{2\sqrt{3}}\right). \tag{IX.2}$$

Howard Georgi, Lie Algebras in Particle Physics: From Isospin to Unified Theories ISBN 0-8053-3153-0

μ^1 is the highest weight of the defining representation (VII.i) generated by the T_a matrices, (VII.1). μ^2 is the highest weight of a different representation. We can find the representation with highest weight μ^2 by a trick, but before we do, let us build all the states by acting on the highest weight state, $|\mu^2>$ with lowering operators. Since $q^1 = 0$, $q^2 = 1$ for this representation, it follows that $\mu^2-\alpha^2$ is a weight but $\mu^2-\alpha^1$ and $\mu^2-2\alpha^2$ are not. Thus, there is a state

$$|\mu^2-\alpha^2> \propto E_{-\alpha^2}|\mu^2>. \tag{IX.3}$$

If now consider acting on $|\mu^2-\alpha^2>$ with lowering operators we see immediately that $E_{-\alpha^2}$ annihilates it since $\mu^2-2\alpha^2$ is not a weight. For $E_{-\alpha^1}$, we compute

$$\frac{2\alpha^1\cdot(\mu^2-\alpha^2)}{\alpha^1\cdot\alpha^1} = q-p = 1. \tag{IX.4}$$

But $p = 0$, because $\mu^2 - \alpha^2+\alpha^1$ is not a weight (Remember, $E_{\alpha^1}E_{-\alpha^2}|\mu^2> = E_{-\alpha^2}E_{\alpha^1}|\mu> = 0$ because $\alpha^1 - \alpha^2$ is not a root). Thus, $q = 1$ and $\mu^2 - \alpha^2 - \alpha^1$ is a weight, corresponding to the state

$$|\mu^2-\alpha^2-\alpha^1> \propto E_{-\alpha^1}E_{-\alpha^2}|\mu^2>. \tag{IX.5}$$

This is the end. By the same sort of argument, you can

show that both $E_{-\alpha^1}$ and $E_{-\alpha^2}$ annihilate (IX.5)

If we plot the three weights, μ^2, $\mu^2-\alpha^2$ and $\mu^2-\alpha^2-\alpha^1$ we get the inverted equilateral triangle shown.

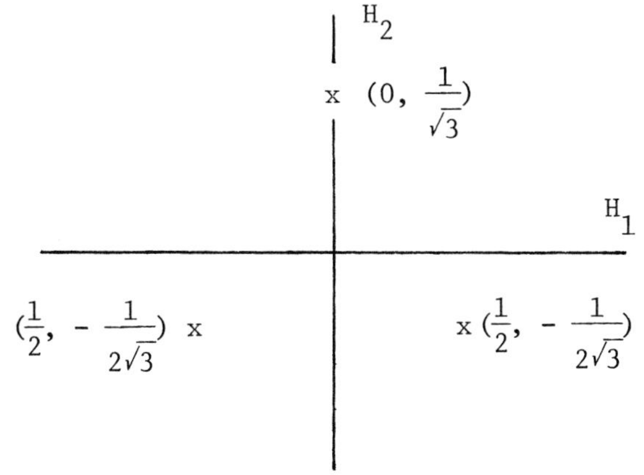

H_2

x $(0, \frac{1}{\sqrt{3}})$

H_1

$(\frac{1}{2}, -\frac{1}{2\sqrt{3}})$ x

x $(\frac{1}{2}, -\frac{1}{2\sqrt{3}})$

(IX.i)

Question: Do all three of the weights correspond to unique states in the representation, or are there some degeneracies which do not show up in our diagram?

We would like to be able to answer this question in general, so let us consider an arbitrary irreducible representation D with highest weight μ. Let $|\mu\rangle$ be any state with weight μ. Then form

$$E_{\phi_1} E_{\phi_2} \cdots E_{\phi_n} |\mu\rangle \tag{IX.6}$$

where ϕ_i are any roots. Obviously, these states, for all n, span the entire irreducible representation. But any such state with a positive ϕ can be dropped without affecting the completeness of the sum. This is because if an E_ϕ for $\phi > 0$ appears, we can move it to the right using the commutation relations, until it acts on $|\mu\rangle$, and thus eliminate it because μ is the highest weight. So we can take all the ϕ to be negative.

But since any negative root is a sum over the simple roots with non-positive integral coefficients, we can take only states of the form

$$E_{-\beta_1} E_{-\beta_2} \cdots E_{-\beta_n} |\mu> \qquad \text{(IX.7)}$$

where the β_i are simple roots, and these states still span the irreducible representation.

This shows that the highest weight corresponds to a unique state, $|\mu>$. Furthermore, any state obtained in a unique way by application of $E_{-\alpha^i}$ is unique. In particular, the states of the form

$$(E_{-\alpha^i})^n |\mu> \qquad \text{(IX.8)}$$

are unique, for any i. This shows that in our example $\mu^2-\alpha^2$ is unique.

To see that the other state is unique, let's figure out exactly what we mean by the obvious symmetry of these representations. It results from the trivial fact that in each root direction, there is an SU(2) group and the representations of SU(2) are symmetric above zero. Formally, if μ is any weight and α any root

$$\mu - \frac{2(\alpha \cdot \mu)}{\alpha^2} \alpha \qquad \text{(IX.9)}$$

is also a weight. This weight is the reflection of the weight μ in the hyperplane perpendicular to α, as shown:

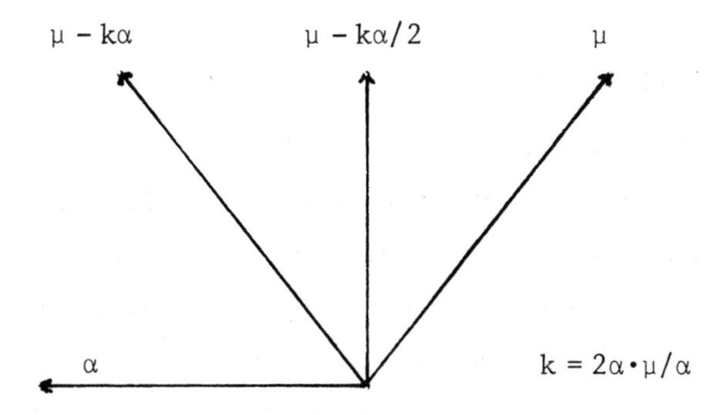

(IX.ii)

Any representation is invariant under all such reflections. The set of all such reflections for all roots forms a group called the Weyl group of the algebra.

Now, back to our example. We know that μ^2 and $\mu^2-\alpha^2$ correspond to unique states. But we can get the weight $\mu^2-\alpha^2-\alpha^1$ by a Weyl reflection of μ^2 in the hyperplane (which in this two dimensional example is just a line) perpendicular to $\alpha^1+\alpha^2$ (note that the root which defines the reflection need not be simple):

$$\mu^2 - \frac{2(\alpha^1\cdot\mu^2+\alpha^2\cdot\mu^2)}{(\alpha^1+\alpha^2)^2}\,(\alpha^1+\alpha^2)$$

$$= \mu^2-\alpha^1-\alpha^2. \qquad \text{(IX.10)}$$

This weight must therefore correspond to a unique state, because its Weyl reflection is unique.

Thus, we have shown that the second fundamental representation, like the first, is three dimensional. Indeed, the weights of the second are just the negatives of the weights of the first, as you can see from (VII.i) and (IX.i). In fact, as mentioned above, the second can be obtained from the first. Here is how.

COMPLEX REPRESENTATIONS

If T_a are the generators of some representation, D, of some Lie algebra, the matrices $-T_a^*$ satisfy the same commutation relations. Thus, the $-T_a^*$ also generate a representation. It is called the complex conjugate of the representation D and is sometimes denoted by $\bar{D}$.

A representation is called real if it is equivalent to its complex conjugate representation. If not, it is complex. If μ is a weight, $-\mu$ is a weight of the complex conjugate representation. This follows because the Cartan generators of the complex conjugate representation are $-H_i^*$. The eigenvalues of H_i^* are the same as those of H_i.

We have seen that a representation is determined by its highest weight. The highest weight of $\bar{D}$ is the negative of the lowest weight of D. Thus, if the lowest weight of D is the negative of the highest weight, D is real. If not, D is complex.

In the defining representation of SU(3), the highest weight is μ^1 but the lowest weight is $-\mu^2$. Thus, its complex conjugate is the second fundamental representation with highest weight μ^2.

We now know that the generators of the second fundamental representation are equivalent to $-T_a^*$, where the T_a are given by (VII.1 and 2).

There are several notations used in physics for SU(3) representations. The most explicit is just to give the ordered pair of integers, (q^1, q^2). But a more common notation, which does not cause too much confusion for small representations is to give their dimension and distinguish between a representation and its complex conjugate (when necessary) with a bar. Thus, (1, 0) = 3 and (0, 1) = $\bar{3}$.

In general, in SU(3), the complex conjugate of the representation with (n, m) is the representation with the q values reversed, (m, n). This is obvious if you think of the representation as built out of fundamental representations, (1, 0) and (0, 1). The highest weight of (n, m) is n times the highest weight of (1, 0) plus m times the highest weight of (0, 1) or $n\mu^1 + m\mu^2$. But the lowest weight of (n, m) is n times the lowest weight of (1, 0) plus m times the lowest weight of (0, 1) or $-n\mu^2 - m\mu^1$. This is just the negative of the highest weight of (m, n).

Representations of the form (n, n) are real.

OTHER REPRESENTATIONS

Let us look at the representation (2, 0), with $q^1=2$, $q^2=0$. It has highest weight

$$2\mu^1 = (1, 1/\sqrt{3}). \qquad \text{(IX.11)}$$

We know that $2\mu^1-\alpha^1$ and $2\mu^1 - 2\alpha^1$ are weights corresponding to unique states, but that $2\mu^1 - \alpha^2$ is not a weight. The Weyl reflections now allow us to write down all the weights as shown:

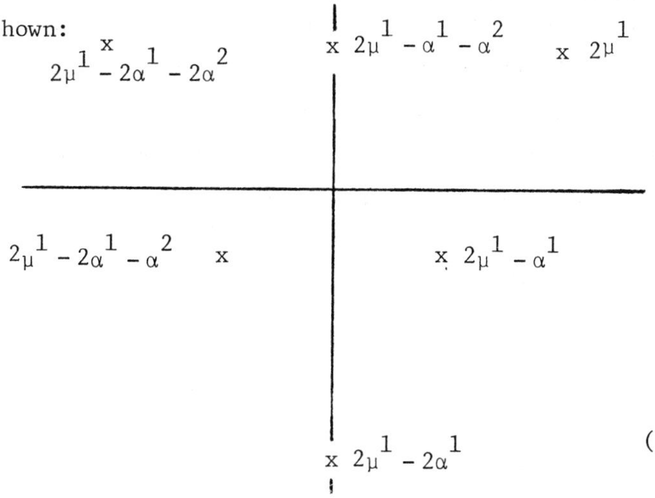

$2\mu^1 - 2\alpha^1 - 2\alpha^2$ x   x $2\mu^1 - \alpha^1 - \alpha^2$ x $2\mu^1$

$2\mu^1 - 2\alpha^1 - \alpha^2$ x   x $2\mu^1 - \alpha^1$

x $2\mu^1 - 2\alpha^1$ (IX.iii)

This is a six-dimensional representation, so $(2, 0) = 6$. Its complex conjugate is $(0, 2) = \bar{6}$.

Next, consider the representation (1, 1) with highest weight $\mu^1+\mu^2$. Notice that we can write

$$\mu^1+\mu^2 = \alpha^1+\alpha^2 . \tag{IX.12}$$

But $\alpha^1+\alpha^2$ is the highest weight of the adjoint representation. Thus, this is the adjoint representation, whose weights are the roots shown in (VII.ii). Thus, $(1, 1) = 8$. Note that the zero weight is doubly degenerate. We can see that directly by noting that there are two ways to get from the highest weight state $|\mu>$ to weight zero with negative simple roots. You can show that the two states,

$$E_{-\alpha^1}E_{-\alpha^2}|\mu> \text{ and} \tag{IX.13a}$$

$$E_{-\alpha^2}E_{-\alpha^1}|\mu>, \tag{IX.13b}$$

are linearly independent (see problem (IX.A)).

Consider the (3, 0) representation. The highest weight is

$$\mu = (\frac{3}{2}, \frac{\sqrt{3}}{2}) = 3\mu^1 .$$

$\mu - m\alpha^1$ for $m = 0$ to 3 and their Weyl reflections correspond to the unique states shown in (IX.iv) on the next page. Obviously, there is a weight in the center. Does it correspond to a unique state? There are two distinct orders of products of $E_{-\alpha^i}$ leading to the same weight

$$E_{-\alpha^1}E_{-\alpha^2}E_{-\alpha^1}|\mu> \quad \text{and} \tag{IX.14a}$$

$$E_{-\alpha^2}E_{-\alpha^1}E_{-\alpha^1}|\mu>. \tag{IX.14b}$$

But the second state, (IX.14b) is

$$[E_{-\alpha^2}, E_{-\alpha^1}]E_{-\alpha^1}|\mu> + E_{-\alpha^1}E_{-\alpha^2}E_{-\alpha^1}|\mu>.$$

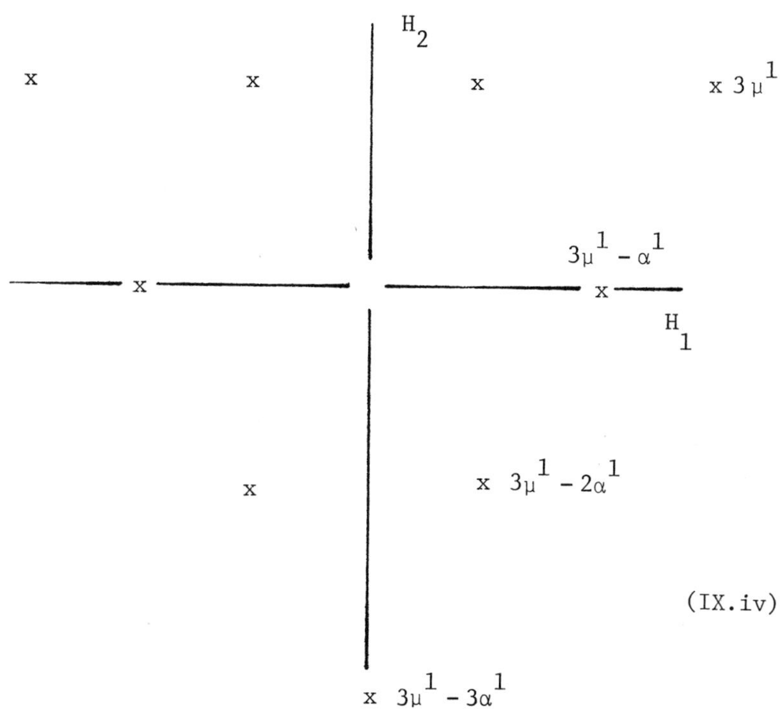

(IX.iv)

The commutator is proportional to $E_{-\alpha^1-\alpha^2}$, thus it commutes with $E_{-\alpha^1}$ because $-2\alpha^1-\alpha^2$ is not a root. So (IX.14b) is

$$E_{-\alpha^1}[E_{-\alpha^2}, E_{-\alpha^1}]|\mu\rangle + E_{-\alpha^1}E_{-\alpha^2}E_{-\alpha^1}|\mu\rangle$$

$$= 2E_{-\alpha^1}E_{-\alpha^2}E_{-\alpha^1}|\mu\rangle$$

because $E_{-\alpha^2}|\mu\rangle = 0$. The two orderings are not linearly independent; so there is a unique state with weight (0, 0). Thus, this is a ten dimensional irreducible representation, $(3, 0) = 10$. Its complex conjugate is $(0, 3) = \overline{10}$.

The weights of a general SU(3) representation form either triangles or hexagons. The triangles are just degenerate hexagons. The representation (2, 1) is shown on the next page. Note that the outer layer of weights is unique, as always. Then, the next layer is doubly degenerate. The general rule is that the degeneracy increases by one unit each time you move in from a hexagonal layer. But once you reach a triangular layer, the degeneracy remains constant as you go in further. Thus, for instance, in the triangular representations, (n, 0) and (0, n), each weight corresponds to a unique state, as in 3, 6, 10 and their complex conjugates.

These facts are easy to prove using the techniques we have already developed. But the proofs are not especially instructive, so we will let them stand without proof. It is also straightforward to find the dimension of an arbitrary representation at this point. But it will be even easier using the results of the next chapter.

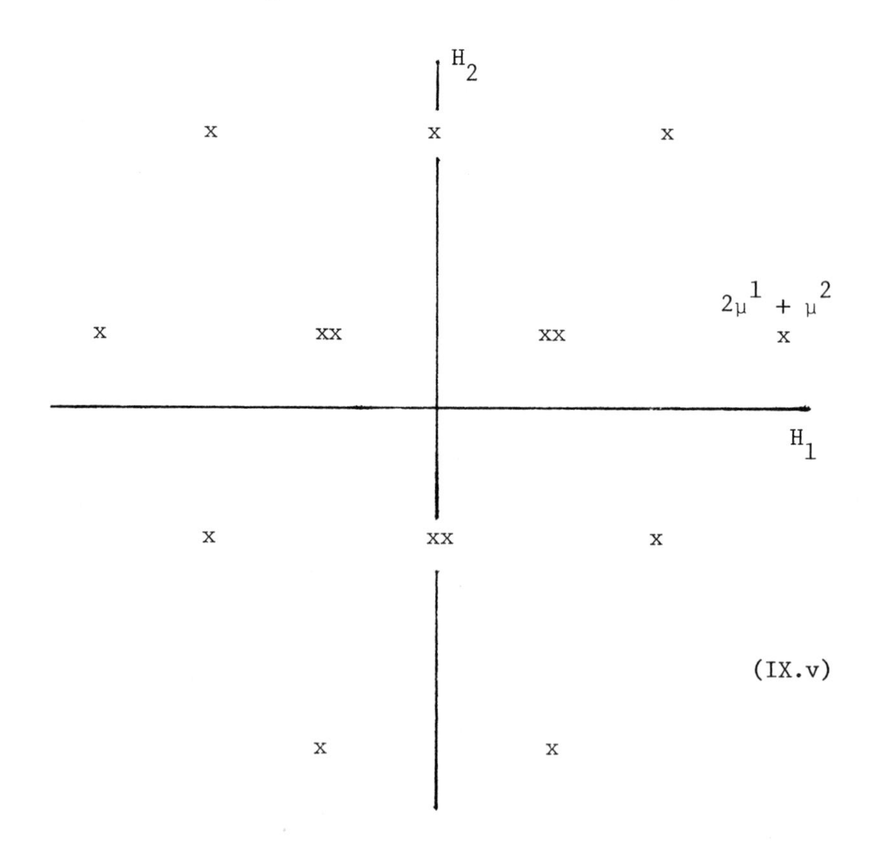

(IX.v)

PROBLEMS FOR CHAPTER IX

(IX.A) If $|\mu\rangle$ is the state of the highest weight ($\mu=\mu^1+\mu^2$) of the adjoint representation of SU(3), show that the states

$$|A\rangle = E_{-\alpha^1}E_{-\alpha^2}|\mu\rangle$$

and

$$|B\rangle = E_{-\alpha^2}E_{-\alpha^1}|\mu\rangle$$

are linearly independent. Hint: Calculate the matrix elements $\langle A|A\rangle$, $\langle A|B\rangle$, $\langle B|A\rangle$ and $\langle B|B\rangle$. Show that $|A\rangle$ and $|B\rangle$ are linearly dependent if and only if

$$\langle A|A\rangle\langle B|B\rangle = \langle A|B\rangle\langle B|A\rangle.$$

(IX.B) Consider the following matrices defined in the six dimensional tensor product space of the SU(3) λ_a matrices and the Pauli matrices σ_a:

$\frac{1}{2}\lambda_a\sigma_2$ for a=1, 3, 4, 6 and 8;

$\frac{1}{2}\lambda_a$ for a=2, 5 and 7.

Show that these generate a reducible representation of SU(3) and reduce it.

(IX.C) Decompose the tensor product of $3 \otimes 3$ using highest weight techniques.

X. TENSOR METHODS

Here we introduce a very useful tool for doing explicit calculations in SU(3) and many other groups. It is the generalization of (III.18). We start by labeling the states of the three representation of SU(3) as follows:

$$\left|\frac{1}{2}, \frac{1}{2\sqrt{3}}\right> \equiv \left|{}_1\right>,$$

$$|-\tfrac{1}{2}, \tfrac{1}{2\sqrt{3}}> \equiv |_2>,$$

$$|0, -\tfrac{1}{\sqrt{3}}> \equiv |_3>, \tag{X.1}$$

The 1, 2 and 3 are suggested by the fact that the eigenvectors of the H_1 and H_2 matrices corresponding to the weights in (X.1) are vectors with a single nonzero entry in the first, second or third position respectively. Note, also, that the indices have been written below the line (for reasons which may not yet be obvious).

If we define a set of representation matrices with one upper and one lower index, as follows:

$$(T_a)^i_j = (\lambda_a)_{ij}/2, \tag{X.2}$$

then the triplet $|_i>$ transforms under the algebra as follows

$$T_a|_i> = |_j>(T_a)^j_i. \tag{X.3}$$

Note that the sum over j involves one upper and one lower index.

Label the states of the $\bar{3}$ as

$$|-\tfrac{1}{2}, -\tfrac{1}{2\sqrt{3}}> \equiv |^1>, \quad |\tfrac{1}{2}, -\tfrac{1}{2\sqrt{3}}> \equiv |^2>,$$
$$|0, \tfrac{1}{\sqrt{3}}> \equiv |^3>. \tag{X.4}$$

Then we have,

$$T_a|^i> = -|^j>(T_a)^i_j. \qquad (X.5)$$

This follows because the $\bar{3}$ is the complex conjugate representation generated by

$$-(T_a^*)^j_i = -(T_a^T)^j_i = -(T_a)^i_j. \qquad (X.6)$$

Now I can define, in the usual way, a state in the tensor product of n 3's and m $\bar{3}$'s.

$$\left|\begin{matrix} i_1\cdots i_m \\ j_1\cdots j_n \end{matrix}\right> = |^{i_1}>\cdots|^{i_m}>|_{j_i}>\cdots|_{j_n}>. \qquad (X.7)$$

It transforms as follows

$$T_a\left|\begin{matrix} i_1\cdots i_m \\ j_1\cdots j_n \end{matrix}\right>$$

$$= \sum_{\ell=1}^{n}\left|\begin{matrix} i_1\cdots i_m \\ j_1\cdots j_{\ell-1}\,k\,j_{\ell+1}\cdots j_n \end{matrix}\right>(T_a)^k_{j_\ell}$$

$$- \sum_{\ell=1}^{m}\left|\begin{matrix} i_1\cdots i_{\ell-1}\,k\,i_{\ell+1}\cdots i_m \\ j_1\cdots j_n \end{matrix}\right>(T_a)^{i_\ell}_k. \qquad (X.8)$$

Let a <u>tensor</u> $|v>$ be any state in this tensor product space:

$$|v> = \left|\begin{matrix} i_1\cdots i_m \\ j_i\cdots j_n \end{matrix}\right> v^{j_1\cdots j_n}_{i_1\cdots i_m} \qquad (X.9)$$

characterized by its tensor <u>components</u>

$$v^{j_1\cdots j_n}_{i_1\cdots i_n}.$$

Then we can think of the action of the generators on $|v>$ as an action on the tensor components as follows:

$$T_a|v> = |T_a v>, \tag{X.10}$$

where

$$(T_a v)^{j_1\cdots j_n}_{i_1\cdots i_m}$$

$$= \sum_{\ell=1}^{n} (T_a)^{j_\ell}_{k}\, v^{j_1\cdots k\cdots j_n}_{i_1\cdots i_m}$$

$$- \sum_{\ell=1}^{m} (T_a)^{k}_{i_\ell}\, v^{j_1\cdots j_n}_{i_1\cdots k\cdots i_m} . \tag{X.11}$$

Now let's pick out the states in the tensor product corresponding to the irreducible representation (n, m). The state with highest weight is

$$\left|\begin{matrix}2\ 2\ 2\cdots \\ 1\ 1\ 1\cdots\end{matrix}\right>. \tag{X.12}$$

It corresponds to the tensor v_H

$$v_H{}^{j_1\cdots j_n}_{i_1\cdots i_m} = \delta_{j_1 1}\delta_{j_2 1}\cdots\delta_{j_n 1}\delta_{i_1 2}\cdots\delta_{i_m 2}.$$

We can now obtain all the states in (n, m) by acting on the tensor v_H with lowering operators. The important point is that v_H has two properties which are preserved by the transformation $v_H \to T_a v_H$. Namely, v_H is symmetric in upper indices and in lower indices and it satisfies

$$\delta^{i_1}_{j_1}\, v_H{}^{j_1 j_2\cdots j_n}_{i_1 i_2\cdots i_m} = 0. \tag{X.13}$$

The δ^i_j is called an <u>invariant tensor</u> because of the trivial fact that $(T_a \delta) = 0$

$$(T_a)^i_k \, \delta^k_j - (T_a)^k_j \, \delta^i_k = 0. \tag{X.14}$$

All states in (n, m) therefore correspond to tensors of this form, symmetric in both upper and in lower indices and traceless. It turns out the correspondence also goes the other way; every such tensor yields a state in (n, m).

Tensors (or more precisely, their components) are nice to work with. We can multiply tensors to obtain tensors with different numbers of components. We can use invariant tensors δ^i_j and also ε_{ijk} and ε^{ijk} (which are invariant because of the tracelessness of T_a). But the nicest thing about them is that the octet representation (1, 1) is just a matrix in the tensor language.

CLEBSCH-GORDAN DECOMPOSITION WITH TENSORS

Let's analyze $3 \otimes 3$ using tensor methods. Say one 3 is the tensor v^i and the other u^j. The tensor product is just the product

$$v^i u^j \tag{X.15}$$

which I can write as

$$\frac{1}{2}\,(v^i u^j + v^j u^i) + \frac{1}{2}\,\varepsilon^{ijk}\varepsilon_{k\ell m} v^\ell u^m. \tag{X.16}$$

By symmetrizing and using the invariant tensor ε, we have explicitly decomposed the product into the sum of a 6, the symmetric tensor

$$\frac{1}{2}\,(v^i u^j + v^j u^i) \tag{X.17}$$

and a $\bar{3}$, the lower index object

$$\varepsilon_{k\ell m} v^\ell u^m. \tag{X.18}$$

Thus, $3 \otimes 3 = 6 \oplus \bar{3}$. In general, any time a tensor is not completely symmetric, we can trade two upper indices for a lower index or two lower indices for an upper index using the ε tensor. Indeed, that is why the irreducible representations correspond to symmetric tensors.

Now consider $3 \otimes \bar{3}$ say v^i (3) and $u_j(\bar{3})$. The product $v^i u_j$ can be written

$$\left\{v^i u_j - \frac{1}{3}\,\delta^i_j v^k u_k\right\} + \frac{1}{3}\,\delta^i_j v^k u_k. \qquad \text{(X.19)}$$

The first term is traceless and corresponds to the 8 representations, while the trivial tensor $v^k u_k$ with no indices corresponds to the trivial representation, (0, 0) or 1. Thus, $3 \otimes \bar{3} = 8 \oplus 1$.

One more for good measure. $3 \otimes 8$ with $v^i(3)$ and $u^j_k(8)$. The product $v^i u^j_k$ can be written in terms of

$$(v^i u^j_k + v^j u^i_k - \frac{1}{4}\,\delta^i_k v^\ell u^j_\ell - \frac{1}{4}\,\delta^j_k v^\ell u^i_\ell), \qquad \text{(X.20)}$$

which is a (2, 1) (or 15 because it's 15 dimensional, see below),

$$\varepsilon_{\ell mn} v^m u^n_k + \varepsilon_{kmn} v^m u^n_\ell \qquad \text{(X.21)}$$

a $\bar{6}$, and

$$v^\ell u^j_\ell, \qquad \text{(X.22)}$$

a 3. Thus, $3 \otimes 8 = 15 \oplus \bar{6} \oplus 3$.

Note that (n-m) mod 3 is conserved in these products, because the invariant tensors have n-m = 0 (mod 3). That is δ^i_j changes the number of both upper and lower indices by one and ε^{ijk} (ε_{ijk}) changes the number of upper (lower) indices by 3. The quantity (n-m) mod 3 is called <u>triality</u>.

So far, we have only considered tensor components of

kets. If we take the bra $\langle v|$ corresponding to the tensor ket $|v\rangle$, we find

$$\langle v| = v^{j_1 \cdots j_n}_{i_1 \cdots i_m}{}^{*} \left\langle {}^{i_1 \cdots i_m}_{j_1 \cdots j_n} \right| . \tag{X.23}$$

But the bra transforms under the algebra with an extra minus sign (see (II.13)). Thus, for example, the triplet $\langle_i|$ transforms as

$$-\langle_i|T_a = -\langle_i|T_a|_j\rangle\langle_j|$$

$$= -(T_a)^i_j \langle_j| . \tag{X.24}$$

Comparing (X.24) with (X.5) we see that the bra with a lower index transforms as if it had an upper index. This is just because of the complex conjugation involved in getting from ket to bra. Similarly, the bra with an upper index transforms as if it had a lower index. This suggests that we define the tensor corresponding to a bra state with the upper and lower indices interchanged. Thus, we say the tensor components of the bra tensor $\langle v|$ are

$$\bar{v}^{i_1 \cdots i_m}_{j_1 \cdots j_n} \equiv v^{j_1 \cdots j_n}_{i_1 \cdots i_m}{}^{*} . \tag{X.25}$$

Then when $\langle v|$ is transformed by $-\langle v|T_a$, $\bar{v}$ is transformed by $T_a \bar{v}$.

As an example, consider the matrix element $\langle u|v\rangle$. For the matrix element to make sense, $|u\rangle$ and $|v\rangle$ must live in the same space and must therefore be tensors of the same kind. From (X.9, 23 and 25) we have

$$\langle u|v\rangle = \bar{u}^{k_1 \cdots k_m}_{\ell_1 \cdots \ell_n} \, v^{j_1 \cdots j_n}_{i_1 \cdots i_m} \tag{X.26}$$

$$\langle {}^{k_1 \cdots k_m}_{\ell_1 \cdots \ell_n} \Big| {}^{i_1 \cdots i_m}_{j_1 \cdots j_n} \rangle$$

$$= \bar{u}^{k_1 \cdots k_m}_{\ell_1 \cdots \ell_n} \; v^{j_1 \cdots j_n}_{i_1 \cdots i_m} \; \delta^{i_1}_{k_1} \cdots \delta^{i_m}_{k_m} \; \delta^{\ell_1}_{j_1} \cdots \delta^{\ell_n}_{j_n}$$

$$= \bar{u}^{i_1 \cdots i_m}_{j_1 \cdots j_n} \; v^{j_1 \cdots j_n}_{i_1 \cdots i_m} \; . \qquad \text{(X.26)}$$

All the indices are repeated and summed over (we say they are contracted, for short)! This is as it should be because the matrix element does not transform at all under the group.

Because tensor operators are closely analogous to states, we can extend the concept of tensors to include the coefficients of tensor operators. Thus, if

$$O^{i_1 \cdots i_m}_{j_1 \cdots j_n} \qquad \text{(X.27)}$$

is an operator which transforms like the state (X.7), and we consider the general linear combination

$$W = w^{j_1 \cdots j_n}_{i_1 \cdots i_m} \; O^{i_1 \cdots i_m}_{j_1 \cdots j_n} , \qquad \text{(X.28)}$$

the set of coefficients, w, behave like a tensor.

DIMENSION OF (n, m)

The dimensionality is simple to compute in the tensor language. First, let's calculate the number of independent tensor components for a symmetric tensor with n upper (or lower indices). Only the number of indices with each value (1, 2 or 3) matters. Thus, the number of independent components is the same as the number of ways of separating n identical objects with two identical partitions which is $(n+2)!/(n!2!)$. A tensor which is symmetric in n upper and m lower indices has

$$\frac{(n+2)!}{n!2} \cdot \frac{(m+2)!}{m!2} = B(n, m) \tag{X.29}$$

independent components. But the trace condition says that a symmetric object with n−1 upper and m−1 lower indices vanishes, which is $B(n-1, m-1)$ conditions. Thus, the symmetric, traceless tensor has dimension $D(n, m) = B(n, m) - B(n-1, m-1)$,

$$D(n, m) = \frac{(n+1)(m+1)(n+m+2)}{2} . \tag{X.30}$$

PROBLEMS FOR CHAPTER X

(X.A) Decompose the product of tensor components $u^i v^{jk}$ where $v^{ij} = v^{ji}$ transforms like a 6.

(X.B) Find the matrix elements $\langle u|T_a|v\rangle$ where $|u\rangle$ and $|v\rangle$ are tensors in the adjoint representation with components, u^i_j and v^i_j.

(X.C) In the 6 of SU(3), for each weight find the corresponding tensor component v^{ij}.

XI. HYPERCHARGE AND STRANGENESS

There are other quantities besides isospin which are conserved by the strong interactions but not by the weak interactions. One of these quantities is called strangeness or S, because the particles which carry it seemed strange when they were discovered. Indeed, they still seem strange today, but for different reasons.

Strangeness is an additive quantum number, like electric charge or baryon number. It is denoted by S.

Howard Georgi, Lie Algebras in Particle Physics: From Isospin to Unified Theories ISBN 0-8053-3153-0

The lightest strange particles are the K mesons: an isospin doublet K^+ and K^0 (where the superscript indicates the electric charge in units of the proton charge) with strangeness +1, and their antiparticles, $\bar{K}^0$ and K^- with strangeness -1.

Because the strong interactions conserve strangeness, these particles are produced in pairs with total strangeness zero in high energy collisions (which take place in less than 10^{-23} seconds, so only the strong interactions have time to operate). For example, K mesons can be produced in the reaction

$$P + P \to P + P + K^+ + K^- . \qquad \text{(XI.1)}$$

The $K^\pm$ produced by (XI.1) are short lived because strangeness is not conserved by the weak interactions. They decay in about 10^{-8} seconds. But as with the $\pi^\pm$, that gives experimental physicists enough time to learn a great deal about them. Like the π's, the K's have spin zero and zero baryon number.

There are also strange baryons (that is, particles with baryon number +1). With $S = -1$, there is an isotriplet Σ^+, Σ^0, Σ^- and an isosinglet Λ^0. With $S = -2$ there is an isodoublet Ξ^0, Ξ^-. All these particles satisfy a relation analogous to (V.15). It is

$$Q = T_3 + Y/2 \qquad \text{(XI.2)}$$

where Y is the hypercharge defined by

$$Y = B + S. \tag{XI.3}$$

Something interesting happens if we plot T_3 versus Y for the particles. For the baryons, we get a hexagon:

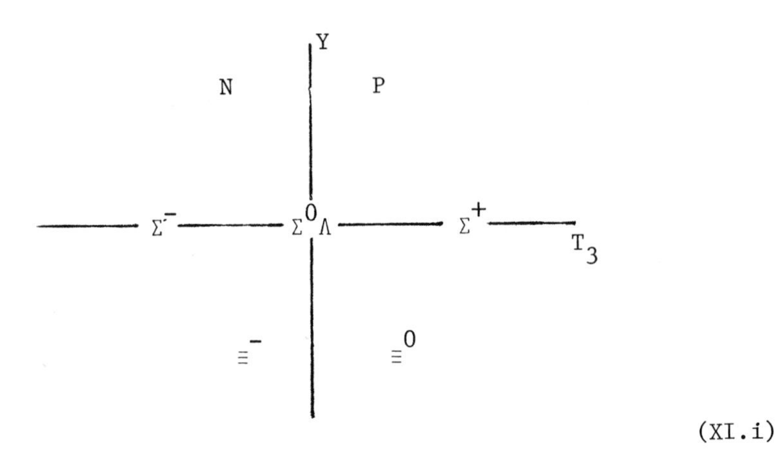

(XI.i)

For the mesons, if we add the η, an isospin singlet with $S = 0$ and a mass close to that of the K's, we get

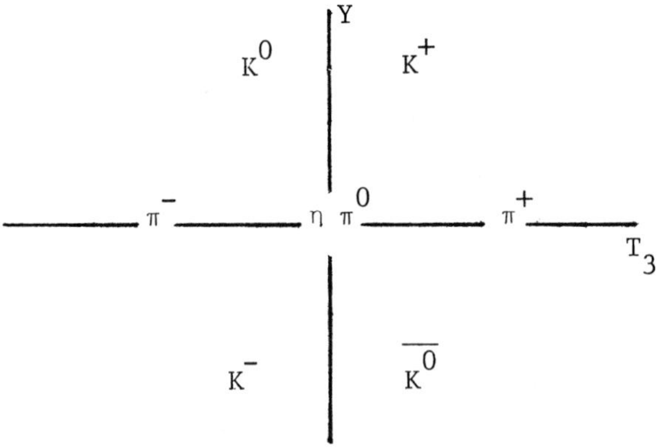

Y

K^0 K^+

π^- $\eta\ \pi^0$ π^+ T_3

K^- $\overline{K^0}$

(XI.ii)

These pictures suggest the 8 of SU(3) with $H_1 = T_3$ and $H_2 = \sqrt{3}\ Y/2$. Can such an SU(3) be an approximate symmetry like isospin? We can divide the strong interactions into two parts, very strong interactions which are invariant under this SU(3) and medium strong interactions which conserve isospin and hypercharge but not the other SU(3) generators. Then if perturbation theory in the medium strong interactions is useful, the particles will fall into SU(3) multiplets. The mass splittings within SU(3) multiplets will be much larger than those within isospin multiplets, but the idea is the same.

THE GELL-MANN OKUBO FORMULA

The pictures (XI.i and ii) are very suggestive, but SU(3) is more than a convenient way of classifying these peculiar particles. We can use it to extract quantitative information. For example, consider the baryon masses.

We can write the baryon states as

$$B^i_j \left| {}^j_i \right> \tag{XI.2}$$

where B^i_j is a tensor that labels the particular baryon. For example a proton state corresponds to $B^i_j = \delta_{i1}\delta_{j3}$. We can exhibit all the states at once by writing the matrix

$$B^i_j = \begin{pmatrix} \frac{\Sigma^0}{\sqrt{2}} + \frac{\Lambda}{\sqrt{6}} & \Sigma^+ & P \\ \Sigma^- & -\frac{\Sigma^0}{\sqrt{2}} + \frac{\Lambda}{\sqrt{6}} & N \\ \Xi^- & \Xi^0 & -\frac{2\Lambda}{\sqrt{6}} \end{pmatrix}_{ij} \tag{XI.3}$$

Now $B^i_j \left| {}^j_i \right> = P|P> + N|N> + \cdots$ The entries of this matrix are convenient labels for keeping track of the states.

We now discuss mass splitting within SU(3) multiplets. If $|p\rangle$ is a single particle state at rest, the mass is (ignoring $H_W + H_{EM}$ and therefore ignoring mass splitting in isospin multiplets)

$$\langle p|H_S|p\rangle = \langle p|H_{VS}|p\rangle + \langle p|H_{MS}|p\rangle \tag{XI.4}$$

where H_{VS} commutes with the SU(3) generators and H_{MS} commutes with isospin (T_i i=1 to 3) and hypercharge (T_8) but not with the others. The first term contributes the same mass to each member of an SU(3) multiplet. In general, we cannot say anything about the contribution of the second term. But suppose that H_{MS} is some component of a tensor operator (the obvious generalization to SU(3)). Then, we can use the Wigner- Eckart theorem to analyze the masses.

In particular, there is reason to believe that H_{MS} transforms like the 8 component of an octet (like the hypercharge, T_8, in other words). That is

$$H_{MS} = (T_8)^i_j \, O^j_i \tag{XI.5}$$

where O^j_i is an octet tensor operator, satisfying

$$[T_a, O^j_i] = O^j_k (T_a)^k_i - O^k_i (T_a)^j_k . \tag{XI.6}$$

We want to know about the matrix elements

$$\langle B|H_{MS}|B\rangle . \tag{XI.7}$$

We can use tensor methods. The matrix element (XI.7) is proportional to the tensors B^i_j, $(T_8)^i_j$ and $\bar{B}^i_j$. But because the matrix element is invariant, all the indices must be contracted. There are only two ways of doing this, we can form the combinations

$$\bar{B}^i_j B^j_k (T_8)^k_i = \mathrm{tr}(B^\dagger B T_8),$$

$$\bar{B}^i_j (T_8)^j_k B^k_i = \mathrm{tr}(B^\dagger T_8 B). \tag{XI.8}$$

We have written (XI.8) in a matrix language in which a tensor with one upper and one lower index is written as a matrix with the rows labeled by the upper index and the columns labeled by the lower index as in (XI.3). In this notation $\bar{B} = B^\dagger$. The notation is particularly convenient for actual calculations because contraction of indices just turns into matrix multiplication and taking traces.

We now know, purely on the basis of the SU(3) symmetry and the assumed transformation law for H_{MS} that the matrix element (XI.7) can be written in terms of two unknown numbers, as follows:

$$\begin{aligned}\langle B|H_{MS}|B\rangle &= X\,\mathrm{tr}(B^\dagger B T_8) + Y\,\mathrm{tr}(B^\dagger T_8 B) \\ &= X(|\Sigma|^2 + |\Xi|^2 - |\Lambda|^2 - 2|N|^2)/12 \\ &\quad + Y(|\Sigma|^2 + |N|^2 - |\Lambda|^2 - 2|\Xi|^2)/12, \qquad \text{(XI.9)}\end{aligned}$$

with a sum over particle types within each isomultiplet understood. Now we can read off the contribution to each particle mass by picking out the appropriate tensor coefficient. Adding a common mass M_0 from H_{VS}, we have (with multiplet names standing for masses)

$$N = M_0 - X/6 + Y/12, \quad \Sigma = M_0 + X/12 + Y/12$$

$$\Lambda = M_0 - X/12 - Y/12, \quad \Xi = M_0 + X/12 - Y/6. \qquad \text{(XI.10)}$$

We have expressed the four masses in terms of three parameters. Symmetry tells us nothing about the values of M_0, X or Y, but we can eliminate them and obtain one relation among the four masses,

$$2(N+\Xi) = 3\Lambda + \Sigma. \qquad \text{(XI.11)}$$

This is the Gell-Mann-Okubo formula. It works! Putting in rounded-off values for the masses, $N = 940$, $\Xi = 1320$,

$\Lambda = 1115$ and $\Sigma = 1190$ MeV, we get agreement within 15 MeV, which is better than we have any right to expect, since splittings within isospin multiplets are of the order of 5 MeV.

There were two crucial steps in the derivation of (XI.10). The first was the guess that H_{MS} transforms like an 8. There is some physics in this guess to which we will return later, but at any rate the success of (XI.10) would seem to vindicate this choice. The second step was (XI.9). This is the generalization to SU(3) of the Wigner-Eckart theorem. In general, whenever we have a matrix element of a tensor operator between tensor states, symmetry requires that all the indices of the coefficients be contracted. Thus, at most, the matrix elements can depend on one dynamical constant for each independent way of contracting all the indices. In more formal group theory language, if we have a matrix element of the form

$$\langle A|B|C\rangle \tag{XI.12}$$

where the state $|A\rangle$ transforms like a representation D_A, the operator B like D_B and the state $|C\rangle$ like D_C (with D_A, D_B and D_C each irreducible), the number of ways of contracting the indices, N, is the number of times the trivial representation appears in the tensor product

$$\bar{D}_A \otimes D_B \otimes D_C . \tag{XI.13}$$

(where the D_A is barred because the bra transforms under the conjugate representation). (XI.13) is obvious if you think of working out the tensor product using tensor methods. The trivial representations are obtained precisely by contracting all the indices. Equivalently, N is the number of times D_A is contained in $D_B \otimes D_C$, because

$D_A \otimes \bar{D}_A$ contains a single trivial representation.

RESONANCES

In πP scattering, there is a large enhancement of the scattering around a fixed mass for the πP system of about 1230 MeV. This is interpreted as the production of a resonance, a particle which decays into πP by strong interactions. The enhancement occurs in all charge states, ++, +, 0, -, which suggests an isospin 3/2 multiplet. This in turn suggests the 10 of SU(3). In fact, all the other particles have been observed.

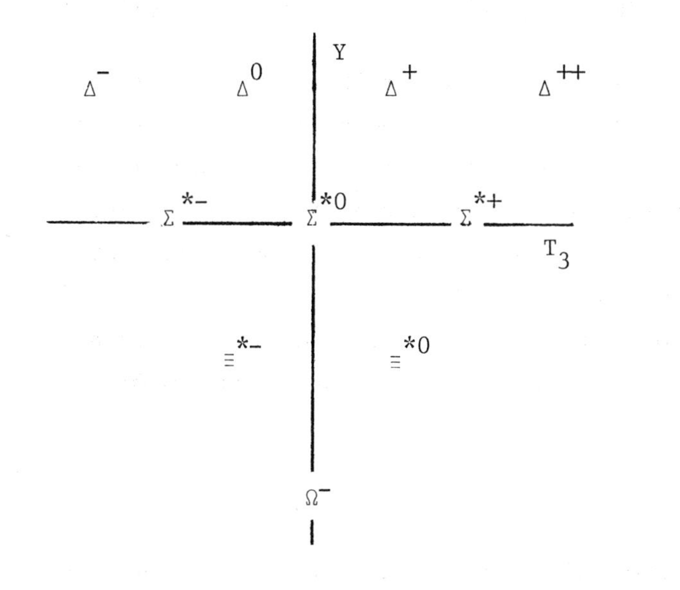

(XI.iii)

When SU(3) was first suggested, the Ω^- had not yet been seen. Gell-Mann was able to predict, not only its existence, but also its mass. Let's repeat the mass prediction.

Assuming as before that H_{MS} transforms like the 8 component of the octet, we need to compute 8×10. This is $(1, 1) \otimes (3, 0) = (4, 1) \oplus (2, 2) \oplus (3, 0) \oplus (1, 1)$. The important thing is that there is a unique 10 in 8×10. So we can compute the matrix element $\langle 10|8|10\rangle$ in terms of a single reduced matrix element: matrix elements of any octet operators are proportional. But we know the matrix elements of one set of octet operators--the group generators themselves. In particular, the matrix element of T_8 is just proportional to hypercharge. Thus, we expect $M = M_0 + kY$, so we predict

$$M_{\Sigma*} - M_{\Delta} = M_{\Xi*} - M_{\Sigma*} = M_{\Omega^-} - M_{\Xi*}. \qquad \text{(XI.14)}$$

Experimentally,

$$M_{\Delta} \simeq 1230 \text{ MeV}, \ M_{\Sigma*} \simeq 1385, \ M_{\Xi*} \simeq 1530. \qquad \text{(XI.15)}$$

The differences between adjacent multiplets is about 150 MeV, so we expect the Ω^- at about 1680. In fact, it was expected to be an almost stable particle, decaying by weak interactions, because it is too light to decay into $K^- \Xi^0$ or $\bar{K}^0 \Xi^-$. The Ω^- has in fact been seen at 1672 MeV decaying weakly into $\Xi\pi$ and ΛK^-.

The success of this prediction convinced almost all particle physicists that SU(3) is a useful approximate symmetry of the strong interactions.

QUARKS

We can think of all these strongly interacting particles as being built out of quarks. Quarks were invented (I suspect) more as a mnemonic device than as a serious prediction. Clearly, since $3 \otimes \bar{3} = 8 \oplus 1$ and $3 \otimes 3 \otimes 3 = 10 \oplus 8 \oplus 8 \oplus 1$, we can build an octet of mesons out of states of one 3 and one $\bar{3}$, and we can build an 8 or 10 of baryons out of state of three 3's (we need an odd number to get spin 1/2). So we can invent three quarks, u, d and s, which form an SU(3) 3.

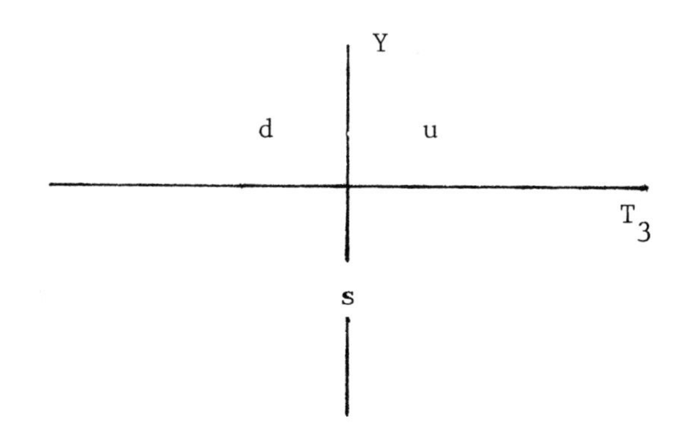

(XI.iv)

The charges are $Q = T_3 + Y/2 = H_1 + H_2/\sqrt{3}$. Thus, the u (up) quark has charge 2/3. The d (down) and s (strange) quarks have charge -1/3.

While no one has ever produced a particle like a quark with fractional charge, we now believe that quarks are more than a mnemonic device. But we will discuss that in detail later.

PROBLEMS FOR CHAPTER XI

(XI.A) What would the Gell-Mann-Okubo argument tell you about the masses of particles transforming like a 6 of SU(3).

(XI.B) Compare the probability for Δ^+ production in $\pi^0 P$ scattering with the probability for Σ^{*0} production in $K^- P$ scattering, assuming SU(3) symmetry for the S matrix.

XII. YOUNG TABLEAUX

We are going to introduce one more notation for the irreducible representations of SU(3), the Young Tableaux. For SU(3) this is not much more than a convenient diagrammatic rewriting of the tensor notation. Its virtues are that it gives an algorithm for the Clebsch-Gordan decomposition of the tensor product of two representations, and that it generalizes to SU(N).

Howard Georgi, Lie Algebras in Particle Physics: From Isospin to Unified Theories ISBN 0-8053-3153-0

The key observation is that the $\bar{3}$ representation is an antisymmetric combination of two 3's, so we can write an arbitrary representation as a tensor product of 3's with appropriate symmetry properties.

Consider the (n, m) representation. It is a tensor (in the old language) whose components are

$$A^{i_1 \cdots i_n}_{j_1 \cdots j_m} \tag{XII.1}$$

symmetric in upper and lower indices and traceless. We can raise the lower indices with ε tensors, obtaining an object with $n+2m$ upper indices

$$a^{i_1 \cdots i_n k_1 \ell_1 \cdots k_m \ell_m} = \varepsilon^{j_1 k_1 \ell_1} \cdots \varepsilon^{j_m k_m \ell_m} A^{i_1 \cdots i_n}_{j_1 \cdots j_m} . \tag{XII.2}$$

Obviously, the a is antisymmetric in each pair $k_x \leftrightarrow \ell_x$ $x = 1$ to m.

Let us associate a box with each index, and arrange the $n+2m$ boxes as follows:

<table>
<tr><td>k_1</td><td>$\cdots$</td><td>k_m</td><td>i_1</td><td>$\cdots$</td><td>i_n</td></tr>
<tr><td>ℓ_1</td><td>$\cdots$</td><td>ℓ_m</td><td></td><td></td><td></td></tr>
</table>

(XII.i)

This is the Young Tableau. To determine the full symmetry of this object, consider the states,

$$a^{i_1 \cdots i_n k_1 \ell_1 \cdots k_m \ell_m} \Big| {}_{i_1 \cdots i_n k_1 \ell_1 \cdots k_m \ell_m} \Big\rangle ,$$

which form the (n, m) representation. Remember, that the state $|_1\rangle$ with weight $(1/2,\ 1/2\sqrt{3})$ has the highest weight in the 3 representation. The next is $|_3\rangle$ with weight $(0,\ -1/\sqrt{3})$. So the highest weight in the (n, m) representation is obtained when all the i's take the value 1, and each k, ℓ pair takes values 1, 3. For the state with highest weight, the tensor component a is non-zero if all the i's and all the k's are 1 and all ℓ's equal 3. The other non-zero components are obtained from this one by antisymmetrization in all the k, ℓ pairs.

All the states of the (n, m) are obtained by acting on this state with lowering operators. The component with all $i = k = 1$ and all $\ell = 3$ goes into a component symmetric in the m+n indices i_1 to i_n and k_1 to k_m, and also symmetric in the m indices ℓ_1 to ℓ_m. The other components are obtained as before by antisymmetrization in each k, ℓ pair.

Comparing with the Young Tableau, we can extract the rule for preparing a tensor with the right symmetry properties to give a state in (n, m). First, symmetrize in the indices in each row of the tableau. Then, antisymmetrize in the indices in each column.

We will be interested in more general tableau with columns of more than two boxes. The rules for forming a tensor are the same as before. Assign an index to each box. Then, symmetrize in the indices in each row and finally antisymmetrize in the indices in each column.

In SU(3), tensors corresponding to tableaux with four or more boxes in any column vanish identically because no tensor can be completely antisymmetric in four or more indices which take on only three values. Any column with three boxes simply yields a factor of the ε tensor in the three corresponding indices

$$\varepsilon^{ijk} = \begin{array}{|c|} \hline i \\ \hline j \\ \hline k \\ \hline \end{array} \tag{XII.ii}$$

So tableaux of the form

$$\begin{array}{|c|c|c|c|c|c|c|c|c|} \hline & \cdots & & k_1 & \cdots & k_m & i_1 & \cdots & i_n \\ \hline & \cdots & & \ell_1 & \cdots & \ell_m \\ \cline{1-6} & \cdots & \\ \cline{1-3} \end{array} \tag{XII.iii}$$

still correspond to the representation (n, m).

With the aid of Young tableaux, we can give an algorithm for the Clebsch-Gordan decomposition of a tensor product. The method is a bit complicated to explain, but stick with it. We'll do lots of examples later.

Suppose we want to decompose the tensor product of irreducible representations α and β corresponding to tableaux A and B. Put a's in the top row of B, b's in the second row. Take the boxes from the top row of B and add them to A, each in a different column, to form new tableaux. Then, take the second row and add them to form tableaux, again each box in a different column; with one additional restriction. Reading from right to left and from the top down, the number of a's must be greater than or equal to the number of b's. This avoids double counting of tensors. The tableaux formed in this way correspond to the irreducible representations in $\alpha \otimes \beta$.

Examples:

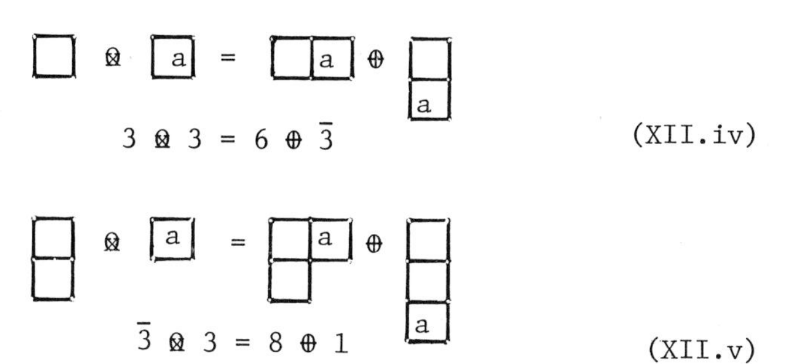

$$3 \otimes 3 = 6 \oplus \bar{3} \qquad \text{(XII.iv)}$$

$$\bar{3} \otimes 3 = 8 \oplus 1 \qquad \text{(XII.v)}$$

A less trivial example is

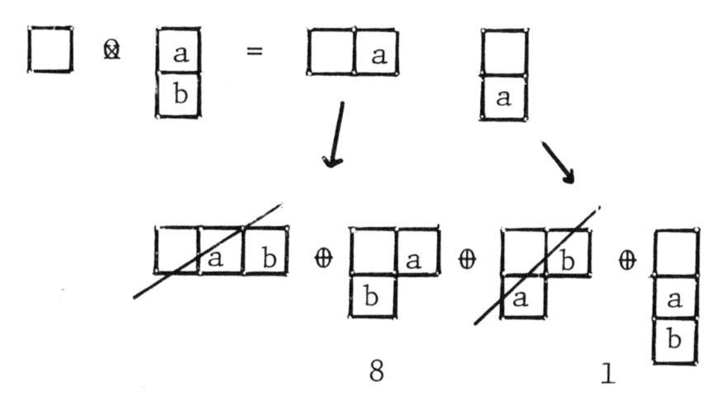

The slashed tableaux do not satisfy the constraint that the number of a's is greater than the number of b's. Note that we did this one in a stupid way (on purpose to illustrate the constraint). A sensible person would work out $\bar{3} \times 3$ and not move so many boxes around as in (XII.v).

Finally, let us work out $8 \otimes 8$;

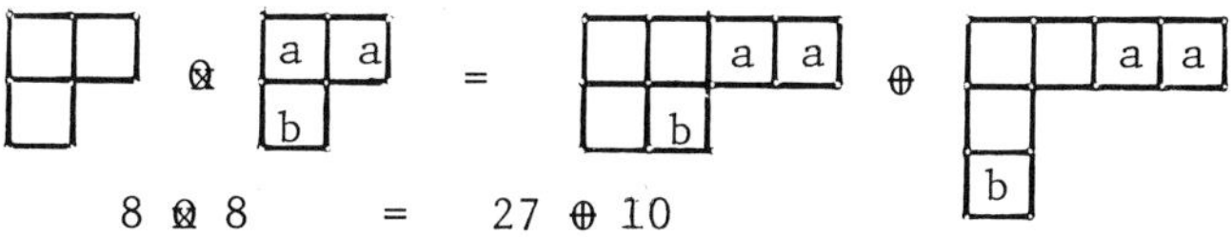

$$8 \otimes 8 = 27 \oplus 10$$

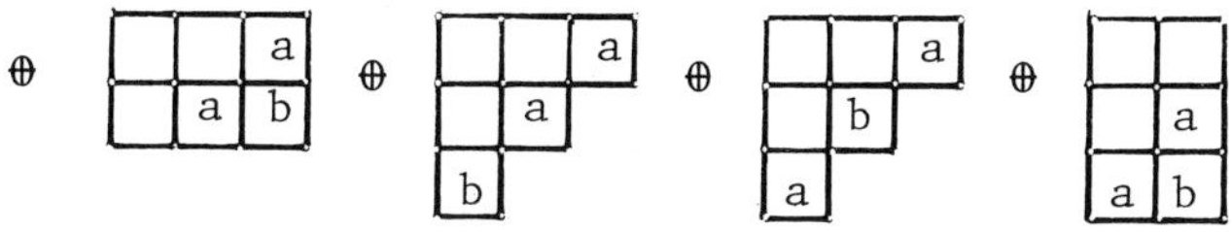

$$\oplus \overline{10} \oplus 8 \oplus 8 \oplus 1 \qquad \text{(XII.vi)}$$

The two 8's in (XII.vi) are different because they have a different pattern of a's and b's.

PROBLEMS FOR CHAPTER XII

(XII.A) Find (2, 1) $\otimes$ (2, 1). Can you determine which representations appear antisymmetrically in the tensor product, and which appear symmetrically?

(XII.B) Find 10 $\otimes$ 8.

(XII.C) You know that for any D, 8 $\otimes$ D must contain D (How do you know?). How can you see this using Young tableaux?

XIII. SU(N)

The group SU(N) of unitary $N \times N$ matrices with det 1 is generated by the hermitian, traceless, $N \times N$ matrices. There are $N^2 - 1$ linearly independent. We will choose a basis in this set satisfying

$$\mathrm{tr}(T_a T_b) = \frac{1}{2}\,\delta_{ab}. \tag{XIII.1}$$

Howard Georgi, Lie Algebras in Particle Physics: From Isospin to Unified Theories ISBN 0-8053-3153-0

We will choose the non-diagonal generators, the E's, to have a single non-zero element each $(1/\sqrt{2})$. The diagonal H's are defined as follows: H_m for $m = 1$ to $N-1$ has m 1's along the diagonal from the upper left-hand corner. The next diagonal element is -m to make it traceless. The rest of the diagonal elements (if any) are zero. Finally, the whole thing is normalized so that (XIII.1) is satisfied. Thus,

$$(H_m)_{ij} = \left(\sum_{k=1}^{m} \delta_{ik}\delta_{jk} - m\,\delta_{i,m+1}\delta_{j,m+1}\right)/\sqrt{2m(m+1)}. \tag{XIII.2}$$

These N^2-1 matrices generate an N dimensional irreducible representation of SU(N) called the defining representation or N for short. The weights of this representation are

$$\nu^1 = \left(\frac{1}{2}, \frac{1}{2\sqrt{3}}, \ldots, \frac{1}{\sqrt{2m(m+1)}}, \ldots, \frac{1}{\sqrt{2(N-1)N}}\right)$$

$$\nu^2 = \left(-\frac{1}{2}, \frac{1}{2\sqrt{3}}, \ldots \qquad \frac{1}{\sqrt{2(N-1)N}}\right)$$

$$\nu^3 = \left(0, -\frac{1}{\sqrt{3}}, \frac{1}{2\sqrt{6}}, \ldots \qquad \frac{1}{\sqrt{2(N-1)N)}}\right)$$

$$\ldots$$

$$\tag{XIII.3}$$

$$\nu^{m+1} = (0,\ 0,\ \cdots,\ 0,\ -\frac{m}{\sqrt{2m(m+1)}},\ \cdots,\ \frac{1}{\sqrt{2(N-1)N}})$$

$$\cdots$$

$$\nu^{N} = (0,\ 0,\ \cdots\cdots\cdots,\ 0,\ \frac{-N+1}{\sqrt{2(N-1)N}})$$

(XIII.3)

For convenience, we choose a backwards convention for positivity of weights. A weight will be called positive if its last non-zero component is positive. With this definition the weights satisfy

$$\nu^1 > \nu^2 > \nu^3 \cdots > \nu^{N-1} > \nu^N. \qquad \text{(XIII.4)}$$

The E matrices take us from any one of these weights to any other. The associated roots are $\nu^i - \nu^j$ for any $i \neq j$. Clearly, the positive roots are $\nu^i - \nu^j$ for $i < j$ and the simple roots are

$$\alpha^i = \nu^i - \nu^{i+1} \quad \text{for} \quad i=1 \text{ to } N-1. \qquad \text{(XIII.5)}$$

These are

$$\alpha^1 = (1,\ 0,\ \cdots\cdots\cdots\ 0)$$

$$\alpha^2 = (-\frac{1}{2},\ \frac{\sqrt{3}}{2},\ 0,\ \cdots\cdots\ 0)$$

$$\alpha^3 = (0,\ -\frac{1}{\sqrt{3}},\ \sqrt{\frac{2}{3}},\ 0,\ \cdots\cdots\ 0)$$

$$\cdots$$

$$\alpha^m = (0,\ 0,\ \cdots\ -\sqrt{\frac{m-1}{2m}},\ \sqrt{\frac{m+1}{2m}},\ 0,\ \cdots\ 0)$$

$$\cdots$$

$$\alpha^{N-1} = (0,\ 0,\ \cdots\cdots,\ -\sqrt{\frac{N-2}{2(N-1)}},\ \sqrt{\frac{N}{2(N-1)}})$$

(XIII.6)

All these roots have length 1. They satisfy

$$\alpha^i \cdot \alpha^{i+1} = -\frac{1}{2}, \quad \alpha^{i^2} = 1,$$

$$\alpha^i \cdot \alpha^j = 0 \quad \text{if} \quad j \neq i \text{ or } i\pm 1. \qquad \text{(XIII.7)}$$

So the Dynkin diagram is

$$\underset{\alpha^1}{\bigcirc} - \underset{\alpha^2}{\bigcirc} \cdots \underset{\alpha^{N-2}}{\bigcirc} - \underset{\alpha^{N-1}}{\bigcirc} \qquad \text{(XIII.i)}$$

Notice that all the weights have the same length and the angles between different weights are the same:

$$|\nu^i|^2 = \frac{N-1}{2N}, \quad \nu^i \cdot \nu^j = -\frac{1}{2N} \quad \text{for } i \neq j. \qquad \text{(XIII.8)}$$

Using these relations, it is easy to check (XIII.7).

But, it is also clear that the weights form a regular figure in N-1 dimensional space, an equilateral triangle for SU(3), a regular tetrahedron for SU(4), and so on. That means that any weight could have been used as a highest weight. And further, all roots are equivalent. But the Dynkin diagram does not treat all roots on the same footing. This is due entirely to the arbitrary ordering we have introduced. If we had invented a different ordering rule, the Dynkin diagram would have the same structure, but the simple roots would be different.

Now let's find the fundamental weights satisfying

$$\frac{2\alpha^i \cdot \mu^j}{\alpha^{i^2}} = \delta^{ij}. \qquad \text{(XIII.9)}$$

It's easy to check that

$$\mu^j = \sum_{k=1}^{j} \nu^k \qquad \text{(XIII.10)}$$

are the fundamental weights.

Just as in SU(3), we can associate states with

tensors,

$$|\nu^i> \rightarrow |_i>, \qquad \text{(XIII.11)}$$

and build up arbitrary representations as tensor products.

In particular, consider the antisymmetric combination of m defining representations. The states are

$$A^{i_1 \cdots i_m} |_{i_1 \cdots i_m}>. \qquad \text{(XIII.12)}$$

The highest weight in this set is obtained when one i=1, another is 2, etc. In other words, the highest weight is

$$\sum_{k=1}^{m} \nu^k = \mu^m. \qquad \text{(XIII.13)}$$

All the antisymmetric states can be obtained by applying lowering operators to the antisymmetric state with highest weight. Thus, the antisymmetric combination of m N's is precisely the fundamental representation with highest weight μ^m.

The highest weight μ of an arbitrary irreducible representation satisfies

$$\mu = \sum_k q_k \mu^k \qquad \text{(XIII.14)}$$

where q_k are non-negative integers. The tensor describing this representation has for each k, q_k sets of k indices, antisymmetric within each set. This suggests that we associate it with the Young tableau on the next page. In fact, this gives a tensor with exactly the right symmetry. Consider the highest weight, which is obtained, for example when all indices in the top row take the value 1, all

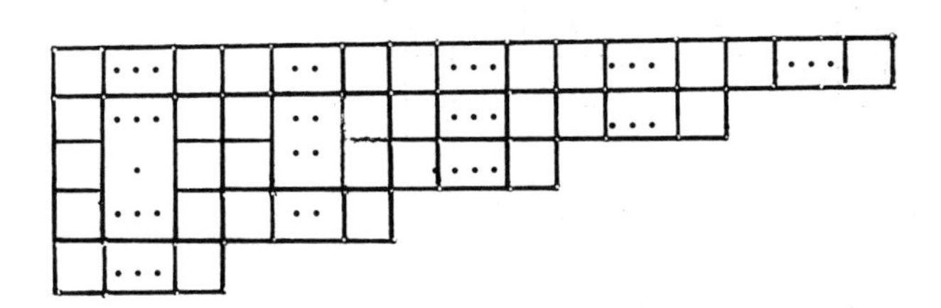

(XIII.ii)

indices in the next row take the value 2, etc. The symmetry of the tableau is the symmetry of the highest weight. First, symmetrize with respect to the indices in each row, then antisymmetrize with respect to the indices in each column.

Just as in SU(3), tableaux with more than N boxes in any column correspond to tensors which vanish identically and any columns with exactly N boxes just contribute a factor of the completely antisymmetric tensor with N indices. Tableaux which are identical except for columns of N boxes correspond to the same irreducible representations.

A useful notation for the irreducible representations of SU(N), based on the Young tableau, is to denote the representation by a nonincreasing series of integers which are just the number of boxes in each column of the Young tableau, $[\ell_1, \ell_2, \cdots]$. Thus, the fundamental representation with highest weight μ^j, which corresponds to a single column of j boxes, is simply [j]. The adjoint representation of SU(N) is [N-1, 1] (the N-1 box column is equivalent to a single lower index).

The rules for Clebsch-Gordan decomposition are now the same as in SU(3). For example, $[1] \otimes [2] = [2, 1] \oplus [3]$ as shown on the next page.

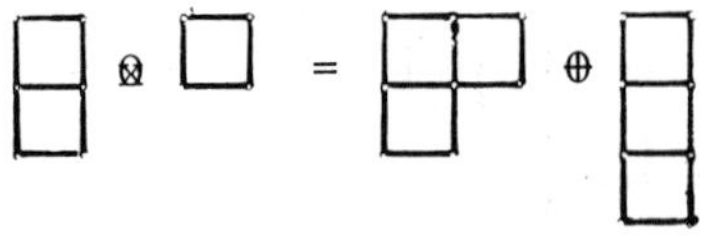

$$[2] \otimes [1] = [2,1] \oplus [3] \qquad \text{(XIII.iii)}$$

DIMENSION OF SU(N) REPRESENTATIONS

For small representations, it is easy to work out the dimension directly in the tensor langauge. For example the dimension of [j] is

$$\binom{N}{j} = \frac{N!}{j!(N-j)!} \,. \qquad \text{(XIII.15)}$$

The dimension of the two column representations can be worked out easily using the Clebsch-Gordan series. For example, it follows from (XIII.iii) that the dimension of [2, 1] is

$$\binom{N}{2} \times \binom{N}{1} - \binom{N}{3} = N(N-1)(N+1)/3. \qquad \text{(XIII.16)}$$

Eventually, though, it gets boring to work out the dimension each time. Fortunately, there is a simple rule for obtaining the dimension of a representation from its tableau. It is called the factors over hooks rule. It works like this. Put a factor of N in the box in the upper left of the tableau. Put integers (factors) in all other boxes, decreasing by one each time you move down and increasing by one each time you move to the right. Call the product of all these integers F. A hook is a line passing vertically up through the bottom of some column of boxes, making a 90° right turn before it reaches the top and passing out through a row of boxes. Call the number of boxes through which the line passes h. Call the product of the h's for all possible hooks H. Then, the dimension of the representation is

$$F/H. \qquad \text{(XIII.17)}$$

For example, for [2, 1], the factors are:

N	N+1
N-1	

(XIII.iv)

The hooks are

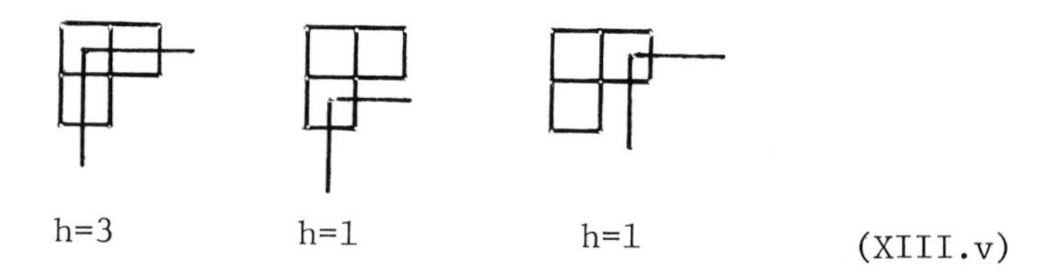

(XIII.v)

Thus, the dimension is

$$F/H = N(N-1)(N+1)/3 \qquad \text{(XIII.18)}$$

in agreement with (XIII.16).

Some of the representations of SU(N) are complex. For example, the lowest weight of the defining representation is ν^N. But, it follows from the tracelessness of the H's that

$$\sum_{j=1}^{N} \nu^j = 0 \qquad \text{(XIII.19)}$$

and thus

$$\nu^N = -\sum_{j=1}^{N-1} \nu^j . \qquad \text{(XIII.20)}$$

This is not $-\nu^1$, hence the representation [1] is certainly complex. In fact, from (XIII.10) it is clear that

$$\overline{[1]} = [N-1]. \qquad \text{(XIII.21)}$$

Similarly, using (XIII.10) and (XIII.19) we see that

$$\overline{[j]} = [N-j]. \qquad \text{(XIII.22)}$$

In general, the complex conjugate of a representation is obtained by replacing each column of j boxes in the tableau with N-j boxes and reading the tableau from right to left. Thus,

$$\overline{[\ell_1, \cdots, \ell_m]} = [N-\ell_m, \cdots, N-\ell_1]. \qquad \text{(XIII.23)}$$

PROBLEMS FOR CHAPTER XIII

(XIII.A) Show that the SU(n) has an SU(n-1) subalgebra. How do the fundamental representations of SU(n) decompose into SU(n-1) representations?

(XIII.B) Find $[3] \otimes [1]$ in SU(5). Check that the dimensions work out.

(XIII.C) Find $[3, 1] \otimes [2, 1]$ in SU(5).

XIV. THE THREE-DIMENSIONAL HARMONIC OSCILLATOR

In the next few chapters, we discuss some applications of SU(N) symmetry to physical systems. The first example may be familiar from introductory quantum mechanics. We discuss the three dimensional harmonic oscillator. Of course, the physics of this simple system is not terribly interesting. But, it will serve as a simple example of the emergence of an SU(N) symmetry and may help you to understand why SU(N) symmetries appear in so many different

Howard Georgi, Lie Algebras in Particle Physics: From Isospin to Unified Theories ISBN 0-8053-3153-0

ways in the later examples.

The Hamiltonian for the 3-dimensional harmonic oscillator is

$$H = \frac{\vec{p}^{\,2}}{2m} + \frac{1}{2} m\omega^2 \vec{x}^{\,2}$$

$$= \hbar\omega(a_k^\dagger a_k + \frac{3}{2}) \tag{XIV.1}$$

where

$$a_k = \sqrt{\frac{m\omega}{2\hbar}}\, x_k + \frac{i}{\sqrt{2m\hbar\omega}}\, p_k ; \quad a_k^\dagger = \sqrt{\frac{m\omega}{2\hbar}}\, x_k - \frac{i}{\sqrt{2m\hbar\omega}}\, p_k \tag{XIV.2}$$

The $a_k (a_k^\dagger)$ are lowering (raising) operators, satisfying

$$[a_k, a_\ell^\dagger] = \delta_{k\ell}$$

$$[a_k^\dagger a_k, a_\ell^\dagger] = a_\ell^\dagger$$

$$[a_k^\dagger a_k, a_\ell] = -a_\ell . \tag{XIV.3}$$

Let $|0\rangle$ be the ground state satisfying

$$a_k|0\rangle = 0 \tag{XIV.4}$$

Then the energy eigenstates are

$$a_{k_1}^\dagger \cdots a_{k_n}^\dagger |0\rangle \tag{XIV.5}$$

with energy $\hbar\omega(n+3/2)$. Notice that the degeneracy of this eigenvalue is the number of symmetric combinations of the n indices $k_1 \cdots k_n$, $(n+2)(n+1)/2$. This is the dimension of the (n, 0) representation of SU(3).

This is not a coincidence! Consider the operators

$$Q_a = a_k^\dagger (T_a)_{k\ell}\, a_\ell \tag{XIV.6}$$

where $T_a = \lambda_a/2$ are the generators of the SU(3) 3.

Obviously, the Q_a generate an SU(3) algebra; since the commutation relations of raising and lowering operators are the same as those of creation and annihilation operators, you have already shown this in problem (V.B). Furthermore,

$$[Q_a, H] = 0, \tag{XIV.7}$$

thus the energy eigenstates fall into degenerate multiplets under SU(3). And

$$Q_a|0\rangle = 0, \tag{XIV.8}$$

thus the ground state is an SU(3) singlet.

The raising and lowering operators are tensor operators under this SU(3):

$$[Q_a, a_k^\dagger] = a_\ell^\dagger (T_a)_{\ell k} \tag{XIV.9}$$

so $a_k^\dagger$ transforms like a 3.

$$\begin{aligned}[Q_a, a_k] &= -(T_a)_{\ell k}\, a_k \\ &= -a_k (T_a^T)_{k\ell} = -a_k (T_a^*)_{k\ell}\end{aligned} \tag{XIV.10}$$

so a_k transforms like a $\bar{3}$.

But because the states are formed by $a^\dagger$'s acting on the (SU(3) invariant) ground state (a's can be eliminated by moving them to the right until they act on $|0\rangle$) and because the $a^\dagger$'s commute among themselves, the state is a symmetric product of 3's. So only representations of the form (n, 0) appear.

This example can easily be extended to SU(N) just by letting k in (XIV.1-6) run from 1 to N, and letting the T_a's be the SU(N) matrices. It should be clear, from this example, why SU(N) arises naturally in a theory with creation and annihilation operators (which satisfy XIV.3 or the like). The number operator $a_k^\dagger a_k$ commutes with the SU(N) generators (XIV.6).

We next describe a rather contrived example of a quantum mechanical system in which all SU(3) (or SU(N)) representations appear in the spectrum.

Consider the following Hamiltonian:

$$\begin{aligned} H = {} & \frac{\omega_1}{2}(\vec{p}^{\,2}+\vec{x}^{\,2}) + \frac{\omega_2}{2}(\vec{P}^2+\vec{X}^2) \\ & + \frac{\Delta}{4}\Big\{(\vec{X}\cdot\vec{x}-P\cdot p)^2 + (\vec{X}\cdot\vec{p}+\vec{x}\cdot\vec{P})^2 \\ & - \vec{x}^{\,2}-\vec{p}^{\,2} - \vec{X}^2-\vec{P}^2\Big\}, \end{aligned} \tag{XIV.11}$$

where the $\vec{X}$ and $\vec{P}$ are an independent set of coordinates and momenta. This describes two harmonic oscillators (with $m_i\omega_i = 1$ for simplicity, and $\hbar = 1$) coupled in what appears to be a very complicated way. Actually it was constructed to be simple. In terms of the raising and lowering operators

$$a_k^\dagger = (x_k - ip_k)/\sqrt{2}, \quad a_k = (x_k + ip_k)/\sqrt{2}$$

$$b_k^\dagger = (X_k - iP_k)/\sqrt{2}, \quad b_k = (X_k + iP_k)/\sqrt{2}, \tag{XIV.12}$$

$$H = \omega_1 (a_k^\dagger a_k + \frac{3}{2}) + \omega_2 (b_k^\dagger b_k + \frac{3}{2}) + \Delta (a_k^\dagger b_k^\dagger)(a_\ell b_\ell). \tag{XIV.13}$$

This Hamiltonian commutes with the SU(3) generators

$$Q_a = a_k^\dagger (T_a)_{k\ell} a_\ell - b_k^\dagger (T_a^*)_{k\ell} b_\ell \tag{XIV.14}$$

Under this SU(3), $a^\dagger(a)$ transforms like a $3(\overline{3})$ as before, but $b^\dagger(b)$ transforms like a $\overline{3}(3)$, so that $(a_\ell b_\ell)$ and $(a_k^\dagger b_k^\dagger)$ are singlets. Now we can construct states out of $a^\dagger$'s and $b^\dagger$'s acting on $|0\rangle$ which transform like an arbitrary SU(3) representation.

This H also commutes with the number operators

$$N_a = a_k^\dagger a_k \quad \text{and} \quad N_b = b_k^\dagger b_k. \tag{XIV.15}$$

The eigenstates of H are thus traceless combinations of n $a^\dagger$'s and m $b^\dagger$'s times singlets

$$(a_{i_1}^\dagger \cdots a_{i_n}^\dagger b_{j_1}^\dagger \cdots b_{j_m}^\dagger - \text{traces}) \cdot (\vec{a}^\dagger \cdot \vec{b}^\dagger)^N |0\rangle \equiv |(n, m)N\rangle, \tag{XIV.16}$$

transforming like the (n, m) representation of SU(3). For example, octet states look like

$$(a_k^\dagger b_\ell^\dagger - \frac{1}{3} \delta_{k\ell} \vec{a}^\dagger \cdot \vec{b}^\dagger)(\vec{a}^\dagger \cdot \vec{b}^\dagger)^N |0\rangle. \tag{XIV.17}$$

The indices of the $a^\dagger$'s act like lower indices under the SU(3) while the indices of the $b^\dagger$'s act like upper indices. The irreducible combinations must be traceless in any pair of one $a^\dagger$ index and one $b^\dagger$ index.

We can calculate the energy eigenvalues by examining any state in each irreducible representation. The most convenient state is the state of highest weight

$$(a_1^\dagger)^n (b_3^\dagger)^m (\vec{a}^\dagger \cdot \vec{b}^\dagger)^N \;|0\rangle = |(n,\ m)N\rangle. \tag{XIV.18}$$

The only term which is nontrivial to calculate is the interaction term proportional to Δ. We can do it as follows:

$$(\vec{a}^\dagger \cdot \vec{b}^\dagger)(\vec{a} \cdot \vec{b})\,|(n,\ m)N\rangle$$

$$= (\vec{a}^\dagger \cdot \vec{b}^\dagger)[\vec{a} \cdot \vec{b},\ \ (a_1^\dagger)^n](b_3^\dagger)^m (\vec{a}^\dagger \cdot \vec{b}^\dagger)^N |0\rangle \tag{XIV.19a}$$

$$+ (\vec{a}^\dagger \cdot \vec{b}^\dagger)(a_1^\dagger)^n [\vec{a} \cdot \vec{b},\ (b_3^\dagger)^m](\vec{a}^\dagger \cdot \vec{b}^\dagger)^N |0\rangle \tag{XIV.19b}$$

$$+ (\vec{a}^\dagger \cdot \vec{b}^\dagger)(a_1^\dagger)^n (b_3^\dagger)^m \;\vec{a} \cdot \vec{b} (\vec{a}^\dagger \cdot \vec{b}^\dagger)^N \;|0\rangle. \tag{XIV.19c}$$

(XIV.19a) is

$$n(\vec{a}^\dagger \cdot \vec{b}^\dagger)(a_1^\dagger)^{n-1} (b_3^\dagger)^m [b_1,\ (\vec{a}^\dagger \cdot \vec{b}^\dagger)^N]\,|0\rangle$$

$$= nN|(n,\ m)N\rangle. \tag{XIV.20}$$

A similar calculation shows that (XIV.19b) is

$$mN|(n,\ m)N\rangle. \tag{XIV.21}$$

(XIV.19c) is

$$(\vec{a}^\dagger \cdot \vec{b}^\dagger)(\vec{a}_1^\dagger)^n (\vec{b}_3^\dagger)^m \;a_k [b_k,\ (\vec{a}^\dagger \cdot \vec{b}^\dagger)^N]\,|0\rangle$$

$$= N(\vec{a}^\dagger \cdot \vec{b}^\dagger)(\vec{a}_1^\dagger)^n (\vec{b}_3^\dagger)^m [\vec{a}_k,\ \vec{a}_k^\dagger (\vec{a}^\dagger \cdot \vec{b}^\dagger)^{N-1}]\,|0\rangle$$

$$= N(N+2)\,|(n,\ m)N\rangle. \tag{XIV.22}$$

Putting (XIV.19–22) together with a straightforward calculation of the free oscillator terms, we find that the H eigenvalue of $|(n, m)N\rangle$ is

$$\omega_1(n + N + \frac{3}{2}) + \omega_2(m + N + \frac{3}{2}) + \Delta N(n + m + N + 2). \tag{XIV.23}$$

This problem is not very interesting as far as physics is concerned, but there are several morals to be drawn from the analysis. Particularly important is the connection between the Δ term, which coupled the a and b operators together and the form of the SU(3) generators (XIV.14). The generators were chosen to make H a singlet. Note, also, the use of the highest weight to calculate the H eigenvalues. The highest weight idea is not just to make the mathematics look nice. This analysis shows that it can simplify explicit calculations.

PROBLEMS FOR CHAPTER XIV

(XIV.A) Show that the operators

$$0^{ij}_k = a^\dagger_i a^\dagger_j a_k - \frac{1}{4}(\delta^i_k a^\dagger_\ell a^\dagger_j a_\ell + \delta^j_k a^\dagger_i a^\dagger_\ell a_\ell)$$

transform like a tensor operator in the (2, 1) representation.

(XIV.B) Calculate the nonzero matrix elements of the operator 0^{11}_3 (where 0^{ij}_k is defined in (XIV.A)) between states of the form

$$a^\dagger_i (a^\dagger_\ell b^\dagger_\ell)|0>$$

and

$$a^\dagger_i a^\dagger_j (a^\dagger_\ell b^\dagger_\ell)|0>.$$

XV. SU(6) AND THE QUARK MODEL

Consider the octet of spin 1/2 baryons and the decuplet of spin 3/2 baryon resonances (and the Ω^-). Typical splittings between octet and decuplet states are not so different from splittings within the SU(3) representations. This suggests that the two representations can be combined into an irreducible representation of some group even larger than SU(3). Since the representations have different spins, the alleged big group cannot commute with angular momentum. You might expect this to cause

Howard Georgi, Lie Algebras in Particle Physics: From Isospin to Unified Theories ISBN 0-8053-3153-0

problems, because it means mixing up internal symmetries and spacetime symmetries. It does. But the problems do not show up until you try to make the theory relativistic, which we will not do.

In the quark model language, the 56 states ($2 \times 8 + 4 \times 10$) correspond to the 56 completely symmetric combinations of three spin 1/2 quarks, each of which comes in $2 \times 3 = 6$ states. The approximate degeneracy of the 8 and 10 suggests that the forces that bind quarks together into hadrons are not only approximately SU(3) invariant but approximately spin independent as well.

All of this suggests that we should look at the group SU(6) where the 6 dimensional representation consists of the 6 quark states. We can label these states in a tensor product notation as products of SU(3) and spin angular momentum basis states:

$$(|u\rangle,\ |d\rangle \text{ or } |s\rangle)\cdot(|\tfrac{1}{2}\rangle \text{ or } |-\tfrac{1}{2}\rangle), \qquad \text{(XV.1)}$$

The generators are products of 3×3 matrices in the SU(3) space and 2×2 matrices in the spin space. In particular, the generators include the SU(3) generators

$$\frac{1}{2}\lambda_a \qquad \text{(XV.2)}$$

where the identity in spin space is understood, and the spin angular momentum generators

$$\frac{1}{2}\sigma_j \qquad \text{(XV.3)}$$

with the identity in SU(3) space understood. In other words, this SU(6) has an SU(3) subalgebra under which the 6 transforms like 2 3's and it has an SU(2) subalgebra under which the 6 transforms like 3 2's. The SU(3) and SU(2) generators commute with one another. In this situation we say that the SU(6) has an SU(3) x SU(2) subalgebra under which the 6 transforms like a (3, 2). The rest of the SU(6) generators are the products

$$\frac{1}{2}\lambda_a\sigma_j \tag{XV.4}$$

for a total of $8+3+8\times 3 = 35$ generators.

The first thing to check is that there really is a 56 dimensional representation with the right properties. The completely symmetric combination of 3 6's is indeed a 56. We can combine the 3 3's and 3 2's both symmetrically into a (10, 4) or we can have antisymmetry in one pair of both SU(3) and SU(2) indices which gives a (8, 2) so

$$(3,\ 2)\otimes(3,\ 2)\otimes(3,\ 2)_{\text{symmetric}} = (10,\ 4)\oplus(8,\ 2) \tag{XV.5}$$

as needed.

Let's see what some of these states look like in the tensor product notation. The decuplet states are simple, symmetric in both SU(3) and SU(2) indices. For example,

$$|\Delta^{++},\ \tfrac{3}{2}\rangle = |uuu\rangle|{+}{+}{+}\rangle$$

$$|\Delta^{+},\ \tfrac{1}{2}\rangle = \frac{1}{3}\{|uud\rangle + |udu\rangle + |duu\rangle\}$$

$$\cdot\{|{+}{+}{-}\rangle + |{+}{-}{+}\rangle + |{-}{+}{+}\rangle\} \tag{XV.6}$$

$$|\Sigma^{*0},\ \tfrac{1}{2}\rangle = \frac{1}{3\sqrt{2}}\{|uds\rangle + 5 \text{ permutations}\}.$$

$$\cdot \{|{+}{+}{-}\rangle + |{+}{-}{+}\rangle + |{-}{+}{+}\rangle\}. \qquad \text{(XV.6)}$$

where we have abbreviated spin ± 1/2 to simply ±. The octet states are more complicated. To get a completely symmetric state, you can multiply an SU(3) state anti-symmetric in a pair of indices by a similarly antisymmetric SU(2) state and then add cyclic permutations to obtain a completely symmetric state, for example

$$|\Lambda, \tfrac{1}{2}\rangle = \frac{1}{2\sqrt{3}} \{(|uds\rangle - |dus\rangle)(|{+}{-}{+}\rangle - |{-}{+}{+}\rangle)$$

$$+ (|sud\rangle - |sdu\rangle)(|{+}{+}{-}\rangle - |{+}{-}{+}\rangle)$$

$$+ (|dsu\rangle - |usd\rangle)(|{-}{+}{+}\rangle - |{+}{+}{-}\rangle)\}. \qquad \text{(XV.7)}$$

Or if it is more convenient (which it usually is), you can multiply an SU(3) state symmetric in a pair of indices with an SU(2) state symmetric in the same pair but with total spin 1/2, then add cyclic permutations. For example,

$$|P, \tfrac{1}{2}\rangle = \frac{1}{3\sqrt{2}} \{|uud\rangle(2|{+}{+}{-}\rangle - |{+}{-}{+}\rangle - |{-}{+}{+}\rangle)$$

$$+ \text{ cyclic permutations}\}.$$

(XV.8)

Now let us use SU(6) symmetry to say something about the magnetic moments of the baryons. To determine the SU(6) properties of a magnetic moment, notice that if a quark is a point particle (similar to an electron) its magnetic moment is

$$\frac{1}{2m} Q \vec{\sigma} \qquad \text{(XV.9)}$$

where m is the quark mass. This is a generator, so we infer that the magnetic moment operator transforms like the 35 dimensional adjoint representation.

We are interested in the matrix elements

$$\langle 56|35|56\rangle. \tag{XV.10}$$

But 35 ⊗ 56 contains a unique 56, so these matrix elements are determined up to an overall factor. They are therefore proportional to the matrix elements of the generator

$$\langle 56|Q\vec{\sigma}|56\rangle. \tag{XV.11}$$

Let us use this to compute the ratio of the P and N moments. Because the states (XV.8) and the neutron state obtained from it by interchanging u and d (and not worrying about an irrelevant overall phase) are σ_3 eigenstates, we must calculate the expectation value of $Q\sigma_3$. The other components vanish. Furthermore, since the baryon states are tensor products, the generator acts on them as a sum over the three quark states. Thus, for example,

$$
\begin{aligned}
&Q\sigma_3|P, \tfrac{1}{2}\rangle = \\
&= \frac{1}{3\sqrt{2}} \Big\{ \frac{2}{3} |uud\rangle(2|{+}{+}{-}\rangle - |{+}{-}{+}\rangle + |{-}{+}{+}\rangle) \\
&+ \frac{2}{3} |uud\rangle(2|{+}{+}{-}\rangle + |{+}{-}{+}\rangle - |{-}{+}{+}\rangle \\
&- \frac{1}{3} |uud\rangle(-2|{+}{+}{-}\rangle - |{+}{-}{+}\rangle - |{-}{+}{+}\rangle) \\
&+ \text{cyclic permutations}\Big\}.
\end{aligned} \tag{XV.12}
$$

Thus, we can calculate

$$
\begin{aligned}
&\langle P, \tfrac{1}{2}|Q\sigma_3|P, \tfrac{1}{2}\rangle = 1, \\
&\langle N, \tfrac{1}{2}|Q\sigma_3|N, \tfrac{1}{2}\rangle = -\frac{2}{3}.
\end{aligned} \tag{XV.13}
$$

So we should have the ratio

$$\frac{\mu_P}{\mu_N} = -\frac{3}{2}. \tag{XV.14}$$

Experimentally (in nuclear magnetons)

$$\mu_P = 2.79, \quad \mu_N = -1.91,$$

$$\frac{\mu_P}{\mu_N} = -1.46. \tag{XV.15}$$

It works. But, notice the tortured logic of the SU(6) calculation. From the quarks, we took only the SU(6) transformation law of the magnetic moment operator. We then used the Wigner-Eckart theorem to show that we can calculate the ratio (XV.14) by calculating the matrix elements of an SU(6) generator.

The quark modeler gets to the answer in one very physical step. The quarks, he says, actually exist. They are not a mathematical device. The baryon is a bound state of three quarks with no angular momentum (indeed, almost at rest). Since the quarks are not orbiting around, the magnetic moment for the baryon is just the sum of the magnetic momentsof the quarks. In the tensor product language, it is just the generator

$$\frac{1}{2m} \sum_{\text{quarks}} Q\vec{\sigma} \tag{XV.16}$$

which is just what we said before, but with one additional piece of information--the scale. If we assume $3m \simeq m_P$,

$$\frac{\mu_P}{2m_P} = \frac{1}{2m} Q\sigma_3 = \frac{1}{2m}$$

thus,

$$\mu_P = m_P/m \simeq 3 \tag{XV.17}$$

which is not bad at all. And, of course, we predict $\mu_N \simeq -2$, which also works rather well.

It looks like we're getting somewhere. The quark model is not only simpler than SU(6), it is more predictive.

Check this idea in another way. Look at μ_Λ.

$$<\Lambda, \tfrac{1}{2}|Q\sigma_3|\Lambda, \tfrac{1}{2}> = -\frac{1}{3} \tag{XV.18}$$

so the SU(6) prediction is

$$\frac{\mu_\Lambda}{\mu_P} = -\frac{1}{3}, \quad \mu_\Lambda = -.93. \tag{XV.19}$$

Experiment gives

$$\mu_\Lambda = -.614 \pm .005. \tag{XV.20}$$

The quark model says

$$\frac{\mu_\Lambda}{2m_P} = \sum_{\text{quark}} \frac{1}{2m_{\text{quark}}} Q\,\sigma_3. \tag{XV.21}$$

But only the strange quark contributes, so in nuclear magnetons,

$$\mu_\Lambda = -\frac{1}{3}\,\frac{m}{m_s}. \tag{XV.22}$$

If we take $m_\Lambda = m_s + 2m$ and $m_P = 3m$,

$$m_s = \frac{3m_\Lambda - 2m_P}{3} = 490 \text{ MeV}, \tag{XV.23}$$

and we obtain

$$\mu_\Lambda = -.64. \tag{XV.24}$$

Thus, the quark model seems to give a good account of some corrections to SU(6) relations.

PROBLEMS FOR CHAPTER XV

(XV.A) Find the SU(6) (i.e. quark model) wave functions for all the spin 1/2 baryon except P, N and Λ (which were discussed in the text).

(XV.B) Use the wave functions you found in (XV.A) to calculate the magnetic moments,

(a) in the SU(6) limit, calculate the ratios to P'

(b) in the quark model, put in SU(3) symmetry breaking by including $m_s \neq m_{u,d}$.

(XV.C) Show that the |Λ, 1/2> state, (XV.7), is an isospin singlet.

XVI. COLOR

There are two things wrong with the simple quark model discussed in the previous chapter. The first is that the connection between spin and statistics seems to break down for the quarks. The quarks must have spin 1/2 in order to produce the spin 1/2 and 3/2 baryons. Thus, we would expect them to obey Fermi-Dirac statistics, like electrons. Furthermore, we would expect the ground state of the three quark system to be symmetric in the exchange

Howard Georgi, Lie Algebras in Particle Physics: From Isospin to Unified Theories ISBN 0-8053-3153-0

of the positions of the three particles. Thus, we would expect the baryon states to be completely antisymmetric in the SU(6) indices of the quarks. This is just the opposite of what is observed. The baryon states fit nicely into the 56 which is completely symmetric in the SU(6) labels.

We could just assume that quarks obey Bose-Einstein statistics (that their wave functions are symmetric) even though they have spin 1/2. But we think that there is something fundamental about the connection between spin and statistics. The theories that we use to describe relativistic quantum mechanics do not make sense for Bose-Einstein spin 1/2 particles.

The second difficulty is that only $q\bar{q}$ and qqq combinations of quarks have been observed. We need some rather peculiar dynamics to explain why these states are seen and all other combinations are not. Certainly, there is not a simple attractive force between quarks, because that would produce qq as well as $q\bar{q}$, for example.

The resolutions of these two difficulties are related and the solution constitutes one of the most fascinating applications of SU(N) to particle physics.

It is not hard to guess that the first problem can be resolved by assuming that the quark has a hidden label, in addition to quark type (which is sometimes called flavor, for short) and spin. If the quark states are completely antisymmetric in the hidden label, then the quarks can be fermions but the ground state will still have the right SU(6) quantum numbers.

We can give a group theoretic significance to antisymmetry in the hidden label if it is an SU(3) index, so that the completely antisymmetric state is an SU(3) singlet.

The SU(3) carried by the quarks is called color. The three SU(3) indices are sometimes called red, green and blue, though we will often be more prosaic and simply call them 1, 2 and 3.

It cannot be emphasized too strongly that color SU(3) has absolutely nothing to do with Gell-Mann's approximate SU(3) which we studied in Chapter XI. Gell-Mann's SU(3) is an approximate symmetry which arises because there are three light quarks, u, d and s. Color SU(3) is, we believe, an exact symmetry which is related to the fact that a baryon is built out of three quarks.

At any rate, if we assume that the quarks, for each spin and type, transform under this color SU(3) like a 3, and further assume that the quarks and antiquarks are bound into color SU(3) singlet states, we get the correct SU(3) and spin assignments for the baryons and for the mesons. The quark wave functions are tensors,

$$q^i, \quad i = 1 \text{ to } 3 \tag{XVI.1}$$

The i is a color index, and we have suppressed flavor, spin and position dependence. We can make a color singlet state out of three quarks by contracting the three upper indices with

$$\varepsilon_{ijk}. \tag{XVI.2}$$

These states are then symmetric under exchange of position, flavor and spin labels. They are the baryons. The antiquarks transform like $\bar{3}$'s under color SU(3) (because they have opposite values of H_1 and H_2 and all other quantum numbers). Their wave functions are tensors with a lower index

$$\bar{q}_i. \tag{XVI.3}$$

We can make color singlet states out of three antiquarks by contracting the three lower indices with

$$\varepsilon^{ijk} . \tag{XVI.4}$$

These are the antibaryons, the antiparticles of the baryons. Finally, we can make color singlet states out of one quark and one antiquark by contracting the upper and lower indices with

$$\delta^i_j . \tag{XVI.5}$$

These are the mesons. In terms of Gell-Mann's SU(3) they are $3 \otimes \bar{3} = 8 \oplus 1$, octets and singlets. In terms of spin SU(2) (assuming that the ground state has no orbital angular momentum), they are $2 \otimes 2 = 3 \oplus 1$, spin one and spin zero. These states include the pseudoscalar meson octet, (XI.ii) and also the spin one octet and singlet mesons shown

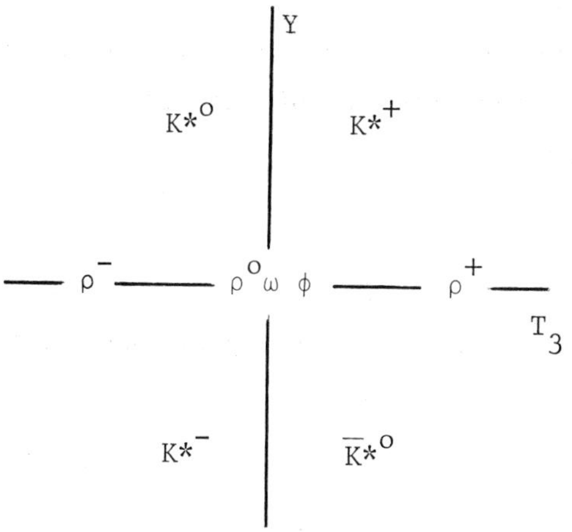

(XVI.i)

Thus, we get one type of state for each of the invariant tensors of SU(3).

We have solved the first problem, but the second remains. Can we find a force which will produce only the color SU(3) singlet states that we want? There is a clue in the nature of the electromagnetic force which is proportional to the product of the charges of the particles involved. We know that this force tends to bind charged particles into neutral atoms. Perhaps if we have a force in which the electromagnetic charges are replaced by the color SU(3) generators, it will tend to bind colored particles into color SU(3) singlets.

To be specific, consider a state of two particles, A and B, transforming according to some representations of color SU(3):

$$T_a^A|r, A\rangle = |s, A\rangle(T_a^A)_{sr},$$
$$T_a^B|x, B\rangle = |y, B\rangle(T_a^B)_{yx}, \tag{XVI.6}$$

where r, s(x,y) label the "colors" in the state A(B). The two particle state is a tensor product,

$$|v, A, B\rangle = v_{rx}|r, A\rangle|x, B\rangle. \tag{XVI.7}$$

We assume that the color force between A and B is proportional to the sum of the products of the color SU(3) generators for A and B,

$$T_a^A T_a^B. \tag{XVI.8}$$

First, let us show that (XVI.8) is consistent with color SU(3) symmetry. The color SU(3) generators for the tensor product space are (from (III.17)).

$$T_a = T_a^A + T_a^B. \tag{XVI.9}$$

These commute with (XVI.8), because

$$\begin{aligned}[T_a^A + T_a^B, T_b^A T_b^B] &= [T_a^A, T_b^A]T_b^B + T_b^A[T_a^B, T_b^B] \\ &= i\, f_{abc}(T_c^A T_b^B + T_b^A T_c^B) = 0.\end{aligned} \tag{XVI.10}$$

Thus, a term like (XVI.8) can appear in the Hamiltonian without spoiling the SU(3) symmetry. Furthermore, the eigenstates of (XVI.8) (and of H) will be irreducible representations of color SU(3).

We can write

$$T_a^A T_a^B = (T_a^2 - T_a^{A2} - T_a^{B2})/2. \tag{XVI.11}$$

The object T_a^2 (summed over a) is called a Casimir operator. It commutes with the SU(3) generators (as in (XVI.10)). Therefore, on each irreducible representation it is just a number (see problem II.D). In some sense, it is a sum of squares of the weights of the representation. It is zero, of course, on a color singlet representation, for which $T_a = 0$. And it increases with increasing size of the representation.

In (XVI.11), T_a^2 is the value of the Casimir operator on the product state, so it depends on how the colors of A and B are combined. T_a^{A2} and T_a^{B2} are fixed. The force between A and B is most attractive when (XVI.11) is most negative (opposite charges attract!). This happens when T_a^2 is as small as it can be. This is just what we need.

In a $q\bar{q}$ state, the most attractive state will clearly be an SU(3) singlet. In a three quark state, we cannot use (XVI.11) directly, but we can consider each two quark subsystem in turn. Each is a qq state. If the colors are combined symmetrically, the product state is a 6. If they are combined antisymmetrically, the product state is a $\bar{3}$. T_a^2 for the $\bar{3}$ is 4/3 while for the 6 it is 10/3, thus the $\bar{3}$ state is most attractive. If all pairs of quarks are in $\bar{3}$'s, then the state is antisymmetric in exchange of any pair of color labels. Thus, it is completely antisymmetric and an SU(3) singlet.

In the quantum theory of electromagnetism (called quantum electrodynamics or QED) the force is associated with the coupling of charged particles to photons. In the quantum theory of the color SU(3) force (called quantum chromodynamics or QCD by analogy) the force is associated with the coupling of colored particles to eight "gluons", one for each SU(3) generator. Despite the close analogy, there are two important differences between QED and QCD. The first is just that in our world, the QCD force is much stronger. The proportionality factors in Coulomb's law are not the same. The second is that the gluons, unlike the photon, carry the charges to which they couple. Because they are associated with the generators, they transform under color SU(3) like the adjoint representation. These facts make the theory more difficult to deal with. We cannot, as yet, extract very precise predictions from QCD as we can from QED. But, there is a great deal of semiquantitative evidence for QCD. For example, the number of colors is measured almost directly (and found to be three) in experiments which measure the probability that a high energy electron-positron collision will produce *hadrons* (a generic word for mesons, baryons and antibaryons--anything built out of quarks and antiquarks).

One conjectured property of QCD which is consistent with everything we know but not completely proven is confinement of quarks and gluons. The conjecture is that the color singlet states are in fact the only states which exist. That free quarks, for example, are impossible. There is some evidence that the force between quark and antiquark becomes constant at large distances, of the order of Λ^2/hc where Λ is a dimensional parameter, a few hundred MeV, which characterizes the strength of the QCD interaction. Thus, free quarks cannot be isolated.

With this picture of the strong interactions, we can understand why SU(3) is a useful symmetry. The interaction of the gluons with the quarks is the same for each flavor. The only distinction between quarks (in the strong interactions) is in their different mass. A mass term in the QCD Hamiltonian looks like

$$m_u u^\dagger u + m_d d^\dagger d + m_s s^\dagger s \qquad \text{(XVI.12)}$$

where u, d, s($u^\dagger$, $d^\dagger$, $s^\dagger$) are creation (annihilation)operators for the u, d and s quarks. (XVI.12) can be rewritten

$$(m_u+m_d+m_s)(u^\dagger u+d^\dagger d+s^\dagger s)/3 \qquad \text{(XVI.13a)}$$

$$+ (m_u-m_d)(u^\dagger u-d^\dagger d)/2 \qquad \text{(XVI.13b)}$$

$$+ (2m_s-m_u-m_d)(2s^\dagger s-u^\dagger u-d^\dagger d)/6. \qquad \text{(XVI.13c)}$$

(XVI.13a) is an SU(3) invariant. (XVI.13b) breaks isospin symmetry. But, we believe that the u-d mass difference is a few MeV, very small compared to the QCD parameter Λ. Thus, this term is a tiny perturbation. It is usually lumped in with the electromagnetic interactions because it is so small, even though it is really a QCD effect. But, at any rate, isospin is a good symmetry.

(XVI.13c) is the term responsible for breaking Gell-Mann's SU(3). It is fairly small because the s quark mass, as measured, for example by the mass splittings within SU(3) multiplets is not large compared to Λ. Notice in particular that (XVI.13c) is a tensor operator. It transforms like T_8. This is the physics behind the Gell-Mann-Okubo formula.

PROBLEMS FOR CHAPTER XVI

(XVI.A) Find a relation between the $q\bar{q}$ force in a meson and the qq force in a baryon.

(XVI.B) Suppose that a "quix", Q, a particle transforming like a 6 under color SU(3) exists. What kinds of bound states would you expect, both of the quix by itself and the quix with ordinary quarks. How would each set of states transform under Gell-Mann's SU(3)?

(XVI.C) Here is a convenient way to calculate the Casimir operators $C(D) = T_a^2$ for small representations, D. Note that

$$\text{tr}\, T_a^2 = \dim(D)C(D) = \sum_a \text{tr}(T_a T_a)$$

$$= \sum_a k_D = 8\, k_D,$$

where k_D is defined in (VI.1) and dim(D) is the dimension of the representation D. $k_D = 1/2$ for D=3 (or $\bar{3}$) so C(3) = 4/3. But k_D behaves in a simple way under $\oplus$ and $\otimes$.

(a) $k_{D_1 \oplus D_2} = k_{D_1} + k_{D_2}$,

(b) $k_{D_1 \otimes D_2} = \dim(D_1)k_{D_2} + \dim(D_2)k_{D_1}$.

Prove (a) and (b) and use them to calculate C(8), C(10), and C(6).

XVII. HADRON MASSES AND HEAVY QUARKS

It would be nice if we could take the QCD theory and use it directly to calculate the meson and baryon mass spectrum. That is too hard. We do not have the tools to deal with such a complicated strongly interacting theory. But, we can use QCD to develop a simple qualitative picture of the hadron masses.

To motivate the discussion, let us consider an imaginary world in which the strong interactions are not so

Howard Georgi, Lie Algebras in Particle Physics: From Isospin to Unified Theories ISBN 0-8053-3153-0

strong. It turns out that we can get from our world to such an imaginary world by turning down the QCD parameter Λ (in our imagination of course). When Λ gets small compared to the masses of the quarks, the most important term in the Hamiltonian is just the quark mass term. We can treat the QCD interaction as a small effect. That should make the strong interactions easier to understand.

If the quarks all had masses very large compared to the QCD parameter Λ, we could do a much better job of predicting hadron masses. In that case, we think that QCD would behave rather like QED, except for the complications due to SU(3) symmetry which we discussed in Chapter XVI. In particular, we could treat the baryons and mesons as nonrelativistic bound states of quarks and antiquarks. The mass of the states would be approximately just the sum of the masses of the constituent quarks. But, there would be small corrections depending on the form of the interaction.

In particular, in this unphysical limit, the approximate SU(6) symmetry discussed in Chapter XV would be easy to understand. The mass term and indeed the leading nonrelativistic interaction term (just the Coulomb interaction) are independent of the spin of the quarks. The most important spin dependent effect for the ground state is the color magnetic moment interaction, a relativistic effect analogous to hyperfine splitting in hydrogen.

There is no good reason to expect this picture to work perfectly for quarks like the u, d and s. But, there are some reasons to expect that it will be qualitatively correct. In fact, in practice, it works rather well. Let us try it.

We will discuss only the ground states in the

baryon sector and the meson sector, and assume that they have zero orbital angular momentum. The states of interest are therefore the SU(6) 56 of baryons comprising the spin 1/2 octet and spin 3/2 decuplet and the $6 \otimes \bar{6}$ of mesons comprising spin 1 and 0 octets and singlets.

Of course, the gross structure of the model works. So long as the mass differences between the quarks are small, we are guaranteed to get approximate Gell-Mann-Okubo formulas for the SU(3) multiplets. The real interest in the model lies in the spin dependent forces. Can we give a good account of the mass differences between states built out of the same quarks but in different spin states, such as Δ^+ and P or Σ^0 and Λ?

The essential physics here is the interaction of magnetic dipoles. Except for the color factor (XVI.11), we assume that the spin dependent force between quarks is just like that between electrons. The dipole moment of the electron or positron is like a current loop. If two current loops are sitting right on top of one another, the state in which the dipole moments are aligned has lower energy than the state in which they are antiparallel.

In a baryon (or meson), the color force between the two quarks (or quark and antiquark) is attractive. Thus, in each pair, the two particles effectively have opposite charges (that sounds peculiar for the baryon, with three pairs, but as we showed in Chapter XVI, color SU(3) really works that way). Therefore, if the magnetic moments are aligned, the spins are paired and vice versa. So, we learn that the state in which spins are paired has lower energy than the state in which they are aligned. Formally, there is a term in the Hamiltonian proportional to

$$\sum_{\substack{\text{pairs}\\ i,j}} \frac{1}{m_i m_j} \vec{\sigma}_i \cdot \vec{\sigma}_j \qquad \text{(XVII.1)}$$

where the σ's are the Pauli matrices acting on the spins. The factors of $1/m_i$ arise because the magnetic moment is inversely proportional to the quark mass.

Evidently, (XVII.1) is what we need to make the decuplet of baryons heavier than the octet, and the spin 1 mesons heavier than the spin 0 mesons. Furthermore, the effect is seen to decrease with the quark mass, as a magnetic moment interaction should. You can see from Tables (XVII.i and ii) at the end of this chapter that

$$\Xi^* - \Xi < \Delta - N, \qquad \text{(XVII.2)}$$

and

$$K^* - K < \rho - \pi \qquad \text{(XVII.3)}$$

(letting particle names stand for their masses). The smaller differences both involve the color magnetic moment of the s quark.

One of the most striking successes of this approach to hadron masses is the explanation of the Σ^0-Λ mass difference. Here not only are the quarks the same, but the total spins of the states are the same. The difference, as you can see from Table (XVII.i), is that in the Λ, the spins of the two light quarks, u-d, are paired while in the Σ^0, they are aligned. Thus, the color magnetic interaction of the light quarks tends to make the Σ^0 heavier. This tendency is partially canceled by the effects of the u-s and d-s pairs because the total spins are the same in the Σ^0 and Λ. But the s quark's moment is smaller by the factor $m_{u,d}/m_s$, thus the light quark pair is more important and the Σ^0 is heavier.

HEAVY QUARKS

Many physicists were first convinced of the utility of the quark model, not by the above analysis of the light hadron masses, but by the prediction and subsequent discovery of heavy quark states exhibiting the spin dependent color magnetic mass splittings. The first clear evidence that nature had actually provided us with heavy quarks to play with came with the discovery of the J/ψ.

The J/ψ is a spin 1 bound state of a heavy quark called c, for charm, with electric charge $Q = 2/3$ and its antiquark $\bar{c}$ (see Table (XVII.ii)). Charmed quarks (indeed any type of quark) are produced only in particle-antiparticle pairs by color SU(3) interactions. Thus, like Y, the charm number C (1 for c and -1 for $\bar{c}$) is conserved by the strong interactions. The c quarks do decay by weak interactions in about 10^{-13} seconds. The J/Ψ state does not have to decay weakly. The c and $\bar{c}$ can simply annihilate into gluons which then turn into light hadrons. But the quark model predicts "charmed" states like $c\bar{u}$, $c\bar{d}$ and cud containing a single charmed quark. The lightest such states must decay weakly.

These states have been observed, and their properties are in excellent agreement with the quark model predictions. First, consider the charmed meson states. The ground state should be spin 0 and spin 1 states of

$$c\bar{u},\ c\bar{d} \text{ and } c\bar{s}. \qquad \text{(XVIII.4)}$$

The spin 0 states are called D^+, D^0 and F^+ and the spin 1 states are D^{*+}, D^{*0} and F^{*+} (see Table (XVII.ii)). These states transform like $\bar{3}$'s under the Gell-Mann SU(3), just like the antiquarks. The c quark is not involved in the Gell-Mann SU(3) at all. However, because the c quark has charge 2/3, (XI.2) is not satisfied. Instead,

$$Q = T_3 + Y/2 + 2C/3. \tag{XVII.5}$$

The D's and D^*'s have been extensively studied. A characteristic prediction of (XVII.1) is that the D^*-D splitting should be much smaller than the ρ-π or K^*-K splitting, because the charmed quark is heavier. Indeed, the D^*-D splitting is about 140 MeV, just as we would expect from (XVII.1).

There should also be charmed baryons, built out of two light quarks and a charmed quark. The lightest of these should also transform like an SU(3) $\bar{3}$, because in the lightest states the spins of the two light quarks will be paired (as in the Λ). Thus, the state will be antisymmetric in Gell-Mann SU(3) indices. At least one of these states, the u d c or Λ_C^+ has been seen (see Table (XVII.i)).

Recently, evidence has been found for the existence of a heavier quark, the b with charge -1/3. The Υ (see Table (XVII.ii)) is presumed to be a bound state of a b and a $\bar{b}$. So far, the states containing single b's have not been extensively studied.

GROUP THEORY VERSUS PHYSICS

Notice that I have resisted the temptation to include the c quark with the u, d and s quarks in the 4 dimensional representation of an SU(4) symmetry. Let me spell out the reasons.

One of the nice things about the QCD quark model is that it explains the success of Gell-Mann's SU(3), because the u, d and s quark mass differences are small compared to the QCD scale parameter Λ. The mass difference between the c quarks and any of the light quarks, on the other hand, is large compared to Λ. SU(4) should not be a useful approximate symmetry.

There is a more general moral here. A symmetry principle should not be an end in itself. Sometimes the physics of a problem is so complicated that symmetry arguments are the only practical means of extracting information about the system. Then, by all means use them. But, do not stop looking for an explicit dynamical scheme that makes more detailed calculation possible. Symmetry is a tool that should be used to determine the underlying dynamics, which must in turn explain the success (or failure) of the symmetry arguments. Group theory is a useful technique, but it is no substitute for physics.

(XVII.i)

	Mass(MeV)	SU(3)	Spin	Ispin	T_3	Y	c
P	938.3	8	$\frac{1}{2}$	$\frac{1}{2}$	$\frac{1}{2}$	1	0
N	939.6	8	$\frac{1}{2}$	$\frac{1}{2}$	$-\frac{1}{2}$	1	0
Λ	1115.6	8	$\frac{1}{2}$	0	0	0	0
Σ^+	1189.4	8	$\frac{1}{2}$	1	1	0	0
Σ^0	1192.5	8	$\frac{1}{2}$	1	0	0	0
Σ^-	1197.3	8	$\frac{1}{2}$	1	-1	0	0
Ξ^0	1314	8	$\frac{1}{2}$	$\frac{1}{2}$	$\frac{1}{2}$	-1	0
Ξ^-	1321	8	$\frac{1}{2}$	$\frac{1}{2}$	$-\frac{1}{2}$	-1	0
Δ^{++}		10	$\frac{3}{2}$	$\frac{3}{2}$	$\frac{3}{2}$	1	0
Δ^+	1230	10	$\frac{3}{2}$	$\frac{3}{2}$	$\frac{1}{2}$	1	0
Δ^0	1234	10	$\frac{3}{2}$	$\frac{3}{2}$	$-\frac{1}{2}$	1	0
Δ^-		10	$\frac{3}{2}$	$\frac{3}{2}$	$-\frac{3}{2}$	1	0
Σ^{*+}	1382	10	$\frac{3}{2}$	1	1	0	0
Σ^{*0}	1382	10	$\frac{3}{2}$	1	0	0	0
Σ^{*-}	1387	10	$\frac{3}{2}$	1	-1	0	0
Ξ^{*0}	1532	10	$\frac{3}{2}$	$\frac{1}{2}$	$\frac{1}{2}$	-1	0
Ξ^{*-}	1535	10	$\frac{3}{2}$	$\frac{1}{2}$	$-\frac{1}{2}$	-1	0
Ω^-	1672	10	$\frac{3}{2}$	0	0	-2	0
Λ_c^+	2273	$\bar{3}$	$\frac{1}{2}$	0	0	$\frac{2}{3}$	1
–	?	$\bar{3}$	$\frac{1}{2}$	$\frac{1}{2}$	$\frac{1}{2}$	$-\frac{1}{3}$	1
–	?	$\bar{3}$	$\frac{1}{2}$	$\frac{1}{2}$	$-\frac{1}{2}$	$-\frac{1}{3}$	1

(XVII.i) cont.

	quark wave function (+ cyclic permutations)
P	$\vert uud\rangle(2\vert{+}{+}{-}\rangle - \vert{+}{-}{+}\rangle - \vert{-}{+}{+}\rangle)/3\sqrt{2}$
N	$-\vert ddu\rangle(2\vert{+}{+}{-}\rangle - \vert{+}{-}{+}\rangle - \vert{-}{+}{+}\rangle)/3\sqrt{2}$
Λ	$(\vert uds\rangle - \vert dus\rangle)(\vert{+}{-}{+}\rangle - \vert{-}{+}{+}\rangle)/2\sqrt{3}$
Σ^+	$\vert uus\rangle(2\vert{+}{+}{-}\rangle - \vert{+}{-}{+}\rangle - \vert{-}{+}{+}\rangle)/3\sqrt{2}$
Σ^0	$(\vert uds\rangle + \vert dus\rangle)(2\vert{+}{+}{-}\rangle - \vert{+}{-}{+}\rangle - \vert{-}{+}{+}\rangle)/6$
Σ^-	$\vert dds\rangle(2\vert{+}{+}{-}\rangle - \vert{+}{-}{+}\rangle - \vert{-}{+}{+}\rangle)/3\sqrt{2}$
Ξ^0	$\vert uss\rangle(\vert{+}{+}{-}\rangle + \vert{+}{-}{+}\rangle - 2\vert{-}{+}{+}\rangle)/3\sqrt{2}$
Ξ^-	$\vert dss\rangle(\vert{+}{+}{-}\rangle + \vert{+}{-}{+}\rangle - 2\vert{-}{+}{+}\rangle)/3\sqrt{2}$
Δ^{++}	$\vert uuu\rangle\vert{+}{+}{+}\rangle$ (no permutations)
Δ^+	$\vert uud\rangle\vert{+}{+}{+}\rangle/\sqrt{3}$
Δ^0	$\vert udd\rangle\vert{+}{+}{+}\rangle/\sqrt{3}$
Δ^-	$\vert ddd\rangle\vert{+}{+}{+}\rangle$ (no permutations)
Σ^{++}	$\vert uus\rangle\vert{+}{+}{+}\rangle/\sqrt{3}$
Σ^{*0}	$(\vert uds\rangle + dus\rangle)\vert{+}{+}{+}\rangle/\sqrt{6}$
Σ^{*-}	$\vert dds\rangle\vert{+}{+}{+}\rangle/\sqrt{3}$
Ξ^{*0}	$\vert uss\rangle\vert{+}{+}{+}\rangle/\sqrt{3}$
Ξ^{*-}	$\vert dss\rangle\vert{+}{+}{+}\rangle/\sqrt{3}$
Ω^-	$\vert sss\rangle\vert{+}{+}{+}\rangle$ (no permutations)
Λ_c^+	$(\vert udc\rangle - \vert duc\rangle)(\vert{+}{-}{+}\rangle - \vert{-}{+}{+}\rangle)/2\sqrt{3}$
-	$(\vert usc\rangle - \vert suc\rangle)(\vert{+}{-}{+}\rangle - \vert{-}{+}{+}\rangle)/2\sqrt{3}$
-	$(\vert dsc\rangle - \vert sdc\rangle)(\vert{+}{-}{+}\rangle - \vert{-}{+}{+}\rangle)/2\sqrt{3}$

(XVII.ii)

	MASS	SU(3)	SPIN	ISPIN	T_3	Y	C	quark wave function
$\pi^{\pm}$	139.6	8	0	1	± 1	0	0	$u\bar{d}(d\bar{u})$
π^0	135.0	8	0	1	0	0	0	$(u\bar{u}-d\bar{d})/\sqrt{2}$
$K^{\pm}$	493.7	8	0	$\frac{1}{2}$	$\pm\frac{1}{2}$	± 1	0	$u\bar{s}(s\bar{u})$
$K^0(\overline{K^0})$	497.7	8	0	$\frac{1}{2}$	$\mp\frac{1}{2}$	± 1	0	$d\bar{s}(s\bar{d})$
η	548	8	0	0	0	0	0	$(2s\bar{s}-u\bar{u}-d\bar{d})/\sqrt{6}$
η'	958	1	0	0	0	0	0	$(u\bar{u}+d\bar{d}+s\bar{s})/\sqrt{3}$
$\rho^{\pm}$	776	1	1	1	± 1	0	0	$u\bar{d}(\bar{d}u)$
ρ^0		1	1	1	0	0	0	$u\bar{d}(\bar{d}u)$
ω	782	$1\oplus 8$	1	0	0	0	0	$(u\bar{u}+d\bar{d})/\sqrt{2}$
$K^{*\pm}$	892	1	0	$\frac{1}{2}$	$\pm\frac{1}{2}$	± 1	0	$u\bar{s}(s\bar{u})$
$K^{*0}(\overline{K^{*0}})$	899	1	1	$\frac{1}{2}$	$\mp\frac{1}{2}$	± 1	0	$d\bar{s}(s\bar{d})$
ϕ	1020	$1\oplus 8$	1	0	0	0	0	$s\bar{s}$
$D^{\pm}$	1868	$\bar{3}(3)$	0	$\frac{1}{2}$	$\pm\frac{1}{2}$	$\frac{1}{3}$	± 1	$c\bar{d}(d\bar{c})$
$D^0(\overline{D^0})$	1863	$\bar{3}(3)$	0	$\frac{1}{2}$	$\frac{1}{2}$	$\mp\frac{1}{3}$	± 1	$c\bar{u}(u\bar{c})$
$F^{\pm}$	-	$\bar{3}(3)$	0	0	0	$\pm\frac{2}{3}$	± 1	$c\bar{s}(s\bar{c})$
$D^{*\pm}$	2009	$\bar{3}(3)$	1	$\frac{1}{2}$	$\frac{1}{2}$	$\mp\frac{1}{3}$	± 1	$c\bar{d}(d\bar{c})$
$D^{*0}(\overline{D^{*0}})$	2006	$\bar{3}(3)$	1	$\frac{1}{2}$	$\frac{1}{2}$	$\mp\frac{1}{3}$	± 1	$c\bar{u}(u\bar{c})$
η_c	2980	1	0	0	0	0	0	$c\bar{c}$
J/ψ	3097	1	1	0	0	0	0	$c\bar{c}$
Υ	9460	1	1	0	0	0	0	$b\bar{b}$

PROBLEMS FOR CHAPTER XVII

(XVII.A) Suppose the quix, Q, described in (XVI.B) is a heavy spin zero particle, so it has no color magnetic moment. Then in the ground state $qq\bar{Q}$ bound states of an antiquix and two u, d or s light quarks, the only spin dependence should come from the light quarks. Discuss the spectrum of all the $qq\bar{Q}$ ground state particles, spin 0 and spin 1, giving their SU(3) properties and an estimate of their masses.

(XVII.B) Estimate $m_{u,d}/m_s$ by comparing ρ-π and K^*-K. Make an independent estimate of the ratio by comparing Δ-N with an appropriate combination of the Σ^*, Σ and Λ masses.

XVIII. UNIFIED THEORIES AND SU(5)

You have seen in Chapter XVI how the strong interactions are associated with a Lie algebra. Most physicists now believe that all of the familiar particle interactions, weak and electromagnetic as well as strong interactions, are associated with Lie algebras in a similar way. This suggests that it may be possible to unify all the particle interactions as different aspects of a single underlying interaction, based on a single simple Lie algebra.

Indeed, we will see that it is possible. In fact, all the particle physics interactions fit remarkably simply

Howard Georgi, Lie Algebras in Particle Physics: From Isospin to Unified Theories ISBN 0-8053-3153-0

into the simple Lie algebra SU(5). The resulting theory is called a grand unified theory, where the term grand is added for obscure historical reasons. We will not be able to discuss the full structure of grand unified theories without the language of quantum field theory. But we can, at least, exhibit the Lie algebraic structure of grand unified theories in detail. I hope this will whet your appetite for a more complete study of the physics behind the group theory.

Let us begin with a (very superficial) introduction to the Glashow-Salam-Weinberg theory of the weak interactions.

One of the salient features of the weak interactions is that they violate parity. Spin 1/2 particles like electrons and quarks, if they are moving, can be characterized by their helicity, the component of the spin in the direction of motion, ± 1/2. Particles with helicity 1/2 are said to be right handed. Those with helicity -1/2 are left handed. The antiparticle of a right-handed particle is left handed and vice versa. Helicity or handedness is clearly not invariant under a parity transformation. A mirror interchanges left and right. Thus, if some interaction acts differently on the right- and left-handed components of a particle, the interaction is parity violating. This is what the weak interactions do.

A massive particle (at least if it can be distinguished from its antiparticle) must have both left- and right-handed components, because the helicity of a massive particle is not a relativistically invariant quantity (it changes sign depending on the frame of reference). But a massless particle need not have both components. So far as we know, the neutrinos are massless. And only left-handed neutrinos and the corresponding antiparticles, right-handed antineutrinos have been observed.

For reasons which will become clear, it is useful to describe the symmetry properties of the creation and annihilation operators, rather than those of the states. We will restrict ourselves to the interactions of the lightest particles, the u and d quarks and the electron and its neutrino. The heavier particles all seem to be copies of one of these.

The Glashow-Salam-Weinberg theory of the weak and electromagnetic interactions treats the creation and annihilation operators as tensor operators under an SU(2) xU(1) Lie algebra, with SU(2) generators R_a and U(1) generator S. Consider the creation operators for the right-handed fields:

$$u^\dagger,\ d^\dagger,\ e^\dagger,\ \bar{u}^\dagger,\ \bar{d}^\dagger,\ \bar{e}^\dagger,\ \bar{\nu}^\dagger, \qquad \text{(XVIII.1)}$$

which create respectively right-handed u quarks, d quarks, electrons, $\bar{u}$ antiquarks, $\bar{d}$ antiquarks, positrons and antineutrinos. The color index on the quark and antiquark fields is suppressed because it plays no role in the weak interactions. Color SU(3) commutes with SU(2) xU(1). Under the SU(2) algebra, the positron and antineutrino and likewise the $\bar{u}$ and $\bar{d}$ antiquarks transform into one another. They are in doublets (because the weak interactions interconvert $\bar{e}$ and $\bar{\nu}$ and $\bar{u}$ and $\bar{d}$):

$$\begin{aligned} &\ell_r^\dagger:\quad \ell_1^\dagger = \bar{e}^\dagger,\ \ell_2^\dagger = \bar{\nu}^\dagger;\\ &\psi_r^\dagger:\quad \psi_1^\ell = \bar{d}^\dagger,\ \psi_2^\dagger = \bar{u}^\dagger. \end{aligned} \qquad \text{(XVIII.2)}$$

Then the commutation relations of the fields (XVIII.1) with the SU(2) and U(1) generators are as follows:

$$[R_a, u^\dagger] = 0, \; [S, u^\dagger] = 2u^\dagger/3;$$

$$[R_a, d^\dagger] = 0, \; [S, d^\dagger] = -d^\dagger/3;$$

$$[R_a, e^\dagger] = 0, \; [S, e^\dagger] = -e^\dagger; \qquad \text{(XVIII.3)}$$

$$[R_a, \psi_r^\dagger] = \psi_s^\dagger \sigma^a_{sr}/2, \; [S, \psi_r^\dagger] = -\psi_r^\dagger/6;$$

$$[R_a, \ell_r^\dagger] = \ell_s^\dagger \sigma^a_{sr}/2, \; [S, \ell_r^\dagger] = \ell_r^\dagger/2.$$

Thus, the fields are tensor operators, with $\psi^\dagger$ and $\ell^\dagger$ being doublets under the SU(2) and the rest singlets.

The annihilation operators for the right-handed states are just the adjoints of (XVIII.1), thus they transform under the complex conjugate representation. The left-handed particles are the antiparticles of these, so their creation operators transform like the annihilation operators of the (corresponding) right-handed states. Thus, for example, the creation operator for a left-handed u quark transforms like the annihilation operator for a right-handed $\bar{u}$ antiquark.

The S values in (XVIII.3) have been constructed so that the electric charge operator Q is

$$Q = R_3 + S. \qquad \text{(XVIII.4)}$$

Then you can check that

$$[Q, u^\dagger] = 2u^\dagger/3, \; [Q, d^\dagger] = -d^\dagger/3,$$

$$[Q, e^\dagger] = e^\dagger, \; [Q, \bar{u}^\dagger] = -2\bar{u}^\dagger/3,$$

$$[Q, \bar{\nu}^\dagger] = 0. \qquad \text{(XVIII.5)}$$

Now the idea is that as in QCD, each of the generators is associated with a particle, the R_a's with three W's and the S with X. One linear combination of W_3 and X is the photon which couples to electric charge Q. Thus, electromagnetism is contained within this larger theory. The other particles, the $W^{\pm}$ (corresponding to the complex combinations $W_1 \pm iW_2$) and the Z (the combination of W_3 and X orthogonal to the photon) are responsible for the weak interactions.

There is something wrong here. $SU(2) \times U(1)$ cannot really be a symmetry. If it were, the weak interactions and electromagnetic interactions would look just the same. In fact, the weak interactions are short range forces which violate parity and change particle identity, while the electromagnetic interactions are long range, conserve parity and do not change particle identities. What picks out one linear combination of the R_3 and X and makes it different from the rest?

The answer is the vacuum. In fact, the $SU(2) \times U(1)$ generators commute with the Hamiltonian, but the vacuum state of our world is not an $SU(2) \times U(1)$ singlet. Hence, we cannot use $SU(2) \times U(1)$ symmetry to conclude, for example, that the $W^{\pm}$ are degenerate with the massless photon. Indeed, there exist states in which there is a massless combination of $W^{\pm}$ and X which acts like the photon, but these states are associated with a different vacuum state. The degenerate particle exists in a different world, not in ours.

This situation is called spontaneous symmetry breakdown. The $SU(2) \times U(1)$ symmetry is said to be spontaneously broken to the U(1) of electromagnetism, because only the linear combination $Q = R_3 + S$ treats the vacuum state of our world as a singlet. The resulting theory gives a very good description of the weak and electromagnetic interactions.

UNIFICATION

The SU(2) x U(1) symmetry (XVIII.3) is a partial unification of the weak and electromagnetic interactions. Might we be able to include the color SU(3) of the strong interactions and further unify the SU(3) and the SU(2) x U(1) (SU(3) x SU(2) x U(1)) into a larger algebra G which is spontaneously broken down to SU(3) x SU(2) x U(1)? To see whether it is possible, let us see how the creation operators transform under SU(3) x SU(2) x U(1). We will use the following convenient notation. We will say that a multiplet of creation operators $a^\dagger_{x,r}$ transforms according to the representation $(D, d)_s$ of SU(3) x SU(2) x U(1) if it satisfies

$$[T_a, a^\dagger_{x,r}] = a^\dagger_{y,r} (T^D_a)_{yx},$$
$$[R_a, a^\dagger_{x,r}] = a^\dagger_{x,s} (R^d_a)_{sr}, \qquad \text{(XVIII.6)}$$
$$[S, a^\dagger_{x,r}] = s\, a^\dagger_{x,r}.$$

Thus, x is a color SU(3) index associated with the SU(3) representation D. The r is an SU(2) index associated with the SU(2) representation d. The S quantum number is s. (XVIII.4) implies that s must be simply the average electromagnetic charge of the multiplet, because for each multiplet,

$$\text{tr } Q = \text{tr } R_3 + \text{tr } S = \text{tr } S. \qquad \text{(XVIII.7)}$$

We know the color SU(3) transformation properties of all the particles, so we can read off the representations of the creation operators for the right-handed particles from (XVIII.3)

$$u^\dagger: \ (3, 1)_{2/3}, \quad d^\dagger: \ (3, 1)_{-1/3}, \quad e^\dagger: \ (1, 1)_{-1},$$

$$\psi^\dagger: \quad (\bar{3}, 2)_{-1/6}, \quad \ell^\dagger: \quad (1, 2)_{1/2}, \tag{XVIII.8}$$

where we have indicated the SU(2) representations by their dimensions, as we do with SU(3) representations. The full SU(3) x SU(2) x U(1) representation of the creation operators for the right-handed particles is thus,

$$(3, 1)_{2/3} \oplus (3, 1)_{-1/3} \oplus (1, 1)_{-1} \oplus (\bar{3}, 2)_{-1/6} \oplus (1, 2)_{1/2}. \tag{XVIII.9}$$

The creation operators for the left-handed fields transform like the complex conjugate of the representation (XVIII.9),

$$(\bar{3}, 1)_{-2/3} \oplus (\bar{3}, 1)_{1/3} \oplus (1, 1)_{1} \oplus (3, 2)_{1/6} \oplus (1, 2)_{-1/2}. \tag{XVIII.10}$$

We have used $\bar{2} = 2$. Notice that the representation is complex because of the parity violating habits of the weak interaction SU(2) x U(1).

(XVIII.9) is the starting point in a search for a unifying algebra. We want to find an algebra G which contains SU(3) x SU(2) x U(1) as a subgroup and which has a representation transforming like (XVIII.9) under this subgroup. Clearly the rank of G must be at least four, if it is to contain the four commuting generators, T_3, T_8, R_3 and S. The simplest possibility is to try the rank four algebra, SU(5).

SU(5) has a five dimensional representation of course, actually two because the 5 and $\bar{5}$ (or [1] and [4]) are not equivalent. Can we find an SU(3) x SU(2) x U(1) subgroup of SU(5) such that the 5 transforms like some five dimensional subset of creation operators (XVIII.9)?

The only possible subset is

$$(3, 1)_{-1/3} \oplus (1, 2)_{1/2}. \qquad \text{(XVIII.11)}$$

The other five dimensional set,

$$(3, 1)_{2/3} \oplus (1, 2)_{1/2}, \qquad \text{(XVIII.12)}$$

is impossible because the generator S is not traceless on it.

It is straightforward to embed SU(3) x SU(2) x U(1) in SU(5) to obtain (XVIII.11). Take the SU(3) to be the traceless generators acting on only the first three indices in the 5 representation and the SU(2) to be the traceless generators acting on the last two:

$$\left(\begin{array}{c|c} T_a & 0 \\ \hline 0 & 0 \end{array}\right), \quad \left(\begin{array}{c|c} 0 & 0 \\ \hline 0 & R_a \end{array}\right). \qquad \text{(XVIII.13)}$$

Then S is

$$\begin{pmatrix} -1/3 & & & & \\ & -1/3 & & 0 & \\ & & -1/3 & & \\ & & & 1/2 & \\ & 0 & & & 1/2 \end{pmatrix}. \qquad \text{(XVIII.14)}$$

Thus, we can put the $d^\dagger$ and $\ell^\dagger$ fields into an SU(5) 5, $\lambda_j^\dagger$, as follows:

$$\lambda_x^\dagger = d_x^\dagger, \quad x = 1, 2 \text{ or } 3;$$
$$\lambda_4^\dagger = \ell_1^\dagger = \bar{e}^\dagger; \quad \lambda_5^\dagger = \ell_2^\dagger = \bar{\nu}^\dagger. \qquad \text{(XVIII.15)}$$

What about the rest of (XVIII.9)? What remains is, $u^\dagger$, $e^\dagger$, and $\psi^\dagger$ which transforms like

$$(3, 1)_{2/3} \oplus (1, 1)_{-1} \oplus (\bar{3}, 2)_{-1/6}. \qquad \text{(XVIII.16)}$$

This representation is 10 dimension. SU(5) has a 10 and a $\overline{10}$, the [2] and [3] representations. The 10 is just an antisymmetric tensor product of two 5's, so we can determine how it transforms under the SU(3) x SU(2) x U(1) subgroup by taking the antisymmetric tensor product of (XVIII.11) with itself. The SU(3) and SU(2) representations compose in the standard way. The S quantum numbers simply add. Thus,

$$[(3, 1)_{-1/3} \oplus (1, 2)_{1/2}] \otimes [(3, 1)_{-1/3} \oplus (1, 2)_{1/2}]_{AS}$$
$$= [(\bar{3}, 1)_{-2/3} \oplus (3,2)_{1/6} \oplus (1, 1)_{1}]. \qquad \text{(XVIII.17)}$$

This is just the complex conjugate of the representation (XVIII.16). Thus, we need the $\overline{10}$ representation, and therefore we can write the remaining creation operators in an SU(5) multiplet antisymmetric in two upper indices, $\chi^{jk\dagger} = -\chi^{kj\dagger}$ with

$$\chi^{ab\dagger} = \varepsilon^{abc} u_c^{\dagger}, \quad a, b, c = 1, 2 \text{ or } 3$$

$$\chi^{a4\dagger} = \psi_2^{a\dagger} = \bar{u}^{a\dagger}, \quad \chi^{a5\dagger} = \psi_1^{a\dagger} = \bar{d}^{a\dagger},$$

$$\chi^{45\dagger} = e^{\dagger}. \qquad \text{(XVIII.18)}$$

This is the standard SU(5) model, with the creation operators for right-handed particles transforming like $5 \oplus \overline{10}$ or equivalently with the creation operators for left-handed particles transforming like the complex conjugate representation $\bar{5} \oplus 10$. The most interesting thing about it is the appearance of quarks, antiquarks and the electron in the same irreducible representation. Because of this, some of the SU(5) interactions do not conserve baryon number. Thus, SU(5) unification leads to proton decay.

PROBLEMS FOR CHAPTER XVIII

(XVIII.A) Find the symmetric tensor product of (XVIII.11) with itself.

(XVIII.B) Do the same for (XVIII.16).

(XVIII.C) Consider the operator

$$0 = \bar{e}^{\dagger} \varepsilon^{abc} u_a u_b d_c$$

where u and d are quark annihilation operators. Show that if the operator 0 appears in the Hamiltonian, it has the right charge and color properties to allow a proton to decay into a π^0 and a positron.

XIX. THE CLASSICAL GROUPS

The set of SU(N) algebras, with Dynkin diagrams

$$\bigcirc\!-\!\bigcirc\cdots\bigcirc\!-\!\bigcirc \qquad \text{(XIX.1)}$$

is one of four infinite sets of Lie algebras which generate what are called the classical groups. The SU(n+1) (rank n) algebras were called A_n by Cartan, who first classified all

Howard Georgi, Lie Algebras in Particle Physics: From Isospin to Unified Theories ISBN 0-8053-3153-0

the simple Lie algebras. In this chapter, we will discuss the other classical groups, before we go on in the next chapter and repeat Cartan's classification. We will discuss these groups very superficially, just enough to determine the Dynkin diagrams. We will return later and get to know each of the groups on a first name basis.

SO(N)

First, consider the group of orthogonal $N \times N$ matrices, which is the group of rotations in an N dimensional real vector space. The generators are the imaginary antisymmetric matrices. If $N = 2n$ is even, we can always divide the 2n dimensional space up into 2 dimensional subspaces, and choose the Cartan generators as follows:

$$(H_m)_{jk} = -i(\delta_{j,2m-1}\delta_{k,2m} - \delta_{k,2m-1}\delta_{j,2m}). \quad \text{(XIX.1)}$$

In other words, H_m is the Pauli matrix σ_2 acting in the m'th subspace. SO(2n) is rank n. The eigenvalues of σ_2 are ± 1, thus the weights of this representation are $\pm e^i$, where e^i are the basis vectors in weight space. The roots move vectors from one two dimensional subspace to another in 4 ways. The corresponding roots are $\pm e^i \pm e^j$ for $i \neq j$. The positive roots are $e^i \pm e^j$ where $i < j$. The simple roots are $e^i - e^{i+1}$, i=1 to n-1 and $e^{n-1} + e^n$. The Dynkin diagram is

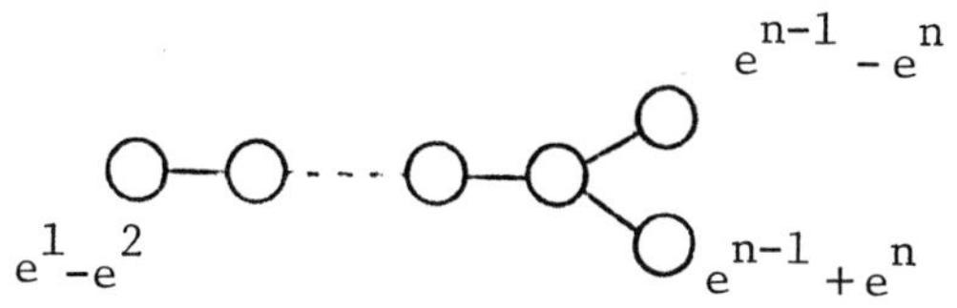

(XIX.ii)

The algebra SO(2n) was called D_n by Cartan.

If $N = 2n+1$ is odd, we can again find n two dimensional subspaces, but there is a one dimensional subspace left over. Define the Cartan subalgebra as before, then the roots corresponding to generators connecting two dimensional subspaces are exactly as before. But there are also roots connecting the one dimensional subspace with the others. These are $\pm e^i$. So the roots are $e^i \pm e^j$, $i \neq j$ and $\pm e^i$. The positive roots are $e^i \pm e^j$ $i < j$ and e^i, and the simple roots are $e^i - e^{i+1}$, i=1 to n-1 and e^n. The Dynkin diagram is

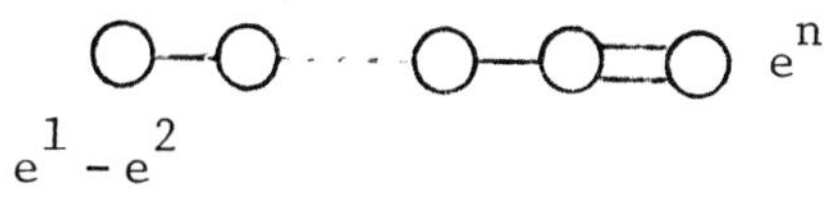

(XIX.iii)

in which e^n is shorter than the rest of the simple roots. These algebras were called B_n by Cartan.

Sp(2n)

Consider the $2n \times 2n$ matrices, which are the following tensor products of 2×2 and $n \times n$ matrices

$$1 \otimes A, \quad \sigma_1 \otimes S_1, \quad \sigma_2 \otimes S_2, \quad \sigma_3 \otimes S_3 \tag{XIX.2}$$

where A is any antisymmetric $n \times n$ matrix and S_1, S_2 and S_3 are arbitrary symmetric $n \times n$ matrices. These form the algebra of the symplectic group Sp(2n). The subset

$$1 \otimes A \quad \text{and} \quad \sigma_3 \otimes S_3 \tag{XIX.3}$$

with traceless S_3 generate an SU(n) subalgebra of matrices of the form

$$\begin{pmatrix} T_a & 0 \\ 0 & -T_a^* \end{pmatrix}. \tag{XIX.4}$$

This is clearly the $n \oplus \bar{n}$ representation. We choose the Cartan subalgebra to contain the Cartan subalgebra of this SU(n), $H_1, H_2 \cdots H_{n-1}$. The final element of the Cartan subalgebra is

$$H_n = \sigma_3 \otimes I/\sqrt{2n}. \tag{XIX.5}$$

The roots in the SU(n) subalgebra have $H_n = 0$, while the first n-1 components are the vectors

$$\nu^i - \nu^j \tag{XIX.6}$$

where the ν^i are the SU(n) weights, (XIII.3). The remaining roots correspond to the matrices

$$(\sigma_1 \pm i\sigma_2) S \tag{XIX.7}$$

where S has the form

$$S_{k\ell} = (\delta_{ik}\delta_{j\ell} + \delta_{i\ell}\delta_{jk}). \tag{XIX.8}$$

Their H_n values are $\pm\sqrt{2/n}$. Their H_m values are $\pm(\nu^i+\nu^j)_m$.

Thus, if ν^{n+1} is a unit vector orthogonal to all the ν^i, $i = 1$ to n, the roots are

$$\nu^i - \nu^j \quad i \neq j \quad \text{and} \quad \pm(\nu^i + \nu^j + \sqrt{\tfrac{2}{n}}\,\nu^{n+1}). \tag{XIX.9}$$

The positive roots are

$$\nu^i - \nu^j, \; i<j \quad \text{and} \quad \nu^i + \nu^j + \sqrt{\tfrac{2}{n}}\,\nu^{n+1}. \tag{XIX.10}$$

The simple roots are

$$\nu^{i} - \nu^{i+1} \quad i=1 \text{ to } n-1 \quad \text{and} \quad 2\nu^{n} + \sqrt{\frac{2}{n}}\,\nu^{n+1} \tag{XIX.11}$$

The Dynkin diagram is

$$\underset{\nu^1-\nu^2}{\bigcirc}\!-\!\bigcirc \cdots \bigcirc\!-\!\bigcirc\!=\!\bigcirc \quad 2\nu^{n} + \sqrt{\frac{2}{n}}\,\nu^{n+1}$$

(XIX.iv)

where the SU(n) roots are shorter. Sp(2n) was called C_n by Cartan.

We will return to all these algebras later and get to know them better. Now we will go on to show that together with five peculiar algebras, these make up all the simple Lie algebras.

PROBLEMS FOR CHAPTER XIX

(XIX.A) Consider the 36 matrices,

$$\sigma_a, \ \tau_a, \ \eta_a, \ \sigma_a\tau_b\eta_c$$

where σ, τ and η are independent Pauli matrices. Show that these matrices form a Lie algebra. Find the roots, the simple roots and the Dynkin diagram. What is the algebra?

(XIX.B) Consider the 28 matrices σ_a, τ_b, η_3, $\sigma_a\eta_1$, $\sigma_a\eta_2$, $\tau_a\eta_1$, $\tau_a\eta_2$, $\sigma_a\tau_b\eta_3$. Show that they form a Lie algebra. Find the roots, the simple roots and the Dynkin diagram. What is the algebra?

XX. THE CLASSIFICATION THEOREM

We have seen that the simple roots of any simple algebra have the following properties:

A. They are linearly independent vectors.

B. If α and β are simple roots, $2\alpha\cdot\beta/\alpha^2$ is a non-positive integer.

To insure that a system of roots yields a simple Lie algebra, we need one additional condition.

C. The simple root system is <u>indecomposable</u>.

A system of roots is said to be <u>decomposable</u> if it can be split into two mutually orthogonal subsystems. A system

Howard Georgi, Lie Algebras in Particle Physics: From Isospin to Unified Theories ISBN 0-8053-3153-0

is indecomposable if it is not decomposable (obviously).

You can show that for decomposable simple-root systems the simple roots in the two orthogonal subsystems commute, and the entire system of roots splits into two commuting subsets. Each subsystem, together with the H operators associated with the subspace it spans, forms an invariant subalgebra. The group associated with a decomposable root system is not simple. It is, however, semi-simple, which means that it has no Abelian invariant subalgebra. Since the subalgebras commute, the subgroups they generate commute. The group G is said to be the direct product of the two subgroups, $G=G_1 x G_2$. All semisimple groups are obtained as direct products of simple groups.

We call a system of vectors satisfying A, B and C a Π-system, following Dynkin. All we have to do to classify the simple Lie algebras is to classify the possible Π-systems, which is simply an exercise in geometry. We will also use "Π-system" to refer to the associated Dynkin diagram. We will now prove some geometrical theorems.

1. The only Π-systems of three vectors are

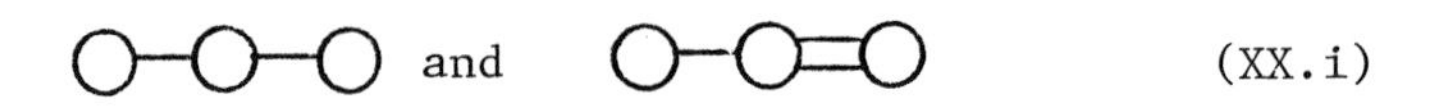

(XX.i)

This follows from the fact that the sum of the angles between any three linearly independent vectors is less than

360°. The only possible angles in a Π-system are 90°, 120°, 135°, and 150°. Only one 90° angle is allowed by indecomposability. So, only these two systems are possible. Note however, that

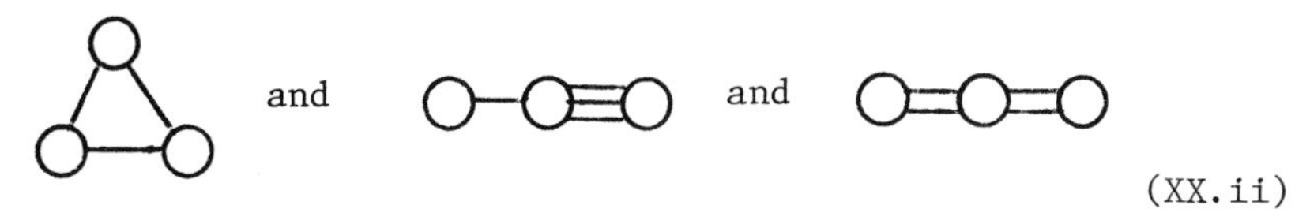

(XX.ii)

satisfy B and C, but because the sum of the angles is 360°, the three vectors are coplanar and not linearly independent.

Obviously, any indecomposable subsystem of a Π-system is another Π-system. So 1. is even stronger. Any three connected (and therefore indecomposable) vectors in a Π-system must be in one of the two configurations in (XX.i).

A trivial corollary is that the only Π-system with a triple line is the rank-2 system

.

(XX.iii)

2. If a Π-system contains two vectors connected by a single line, the diagram obtained by shrinking the line away and merging the two vectors into a single circle is another Π-system.

Let α and β be the two vectors and Γ the set of all the other vectors in the Π-system. Note that Γ contains no vector connected to both α and β (because of 1). The vector $\alpha+\beta$ has the same length as α and β. If $\gamma \in \Gamma$ is a vector connected to α, then $\gamma \cdot (\alpha+\beta) = \gamma \cdot \alpha$. Similarly, if $\gamma' \in \Gamma$ is connected to β, $\gamma' \cdot (\alpha+\beta) = \gamma' \cdot \beta$. So the set $\alpha+\beta$ and Γ is the Π-system described in 2.

This has two immediate corollaries: No Π-system contains more than one double line; and no Π-system contains a closed loop. Either configuration can be shrunk into conflict with 1.

3. If the configuration

(XX.iv)

is a Π-system, for some subdiagram A, then

(XX.v)

is another Π-system.

To see this, label the vectors as follows:

(XX.vi)

We know that $\alpha \cdot \beta = 0$ and

$$2\alpha \cdot \gamma/\alpha^2 = 2\alpha \cdot \gamma/\gamma^2 = 2\beta \cdot \gamma/\beta^2 = 2\beta \cdot \gamma/\gamma^2 = -1. \quad \text{(XX.1)}$$

It follows that

$$2\gamma \cdot (\alpha+\beta)\gamma^2 = -2, \quad 2\gamma \cdot (\alpha+\beta)/(\alpha+\beta)^2 = -1. \quad \text{(XX.2)}$$

Thus,

(XX.vii)

is a Π-system.

A corollary is that the only branches in a Π-system have the form

(XX.viii)

because

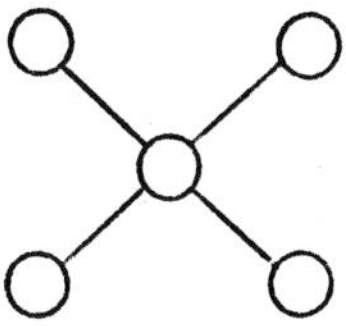

(XX.ix)

can be shrunk by 3 to [diagram] which is not a Π-system according to 1.

Another obvious corollary is that no Π-system contains both a branch point and a double line, or two branches.

This is as far as we can go without considering some peculiar special cases. They are associated, we will see later, with the so-called exceptional groups.

4. No Π-system contains any of the following diagrams:

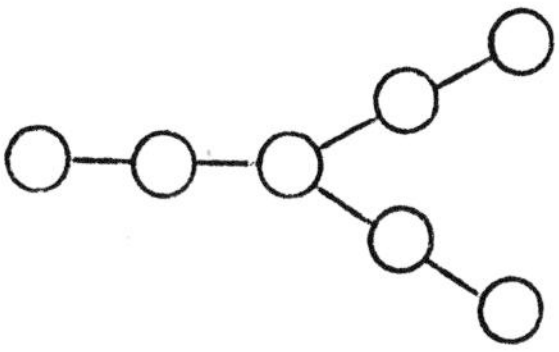

(XX.xa)

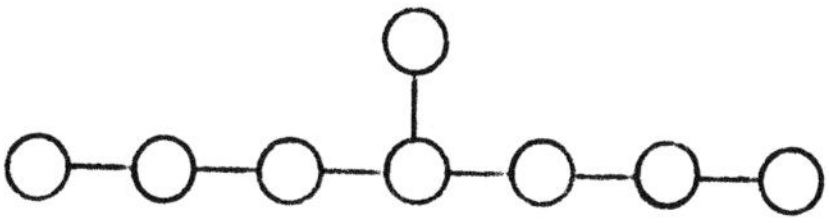

(XX.xb)

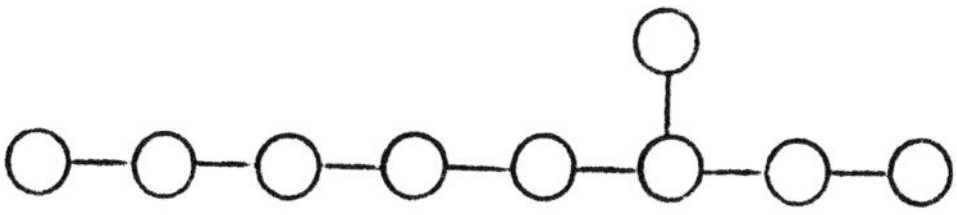

(XX.xc)

(XX.xd)

Assume that we can find systems of vectors with the angles shown in these diagrams (if not 4 is clearly true). We will show that they cannot be linearly independent.

If α^i are the vectors in one of these diagrams, we can find constants μ_i such that

$$(\sum_i \mu_i \alpha^i)^2 = 0. \qquad \text{(XX.3)}$$

For a.

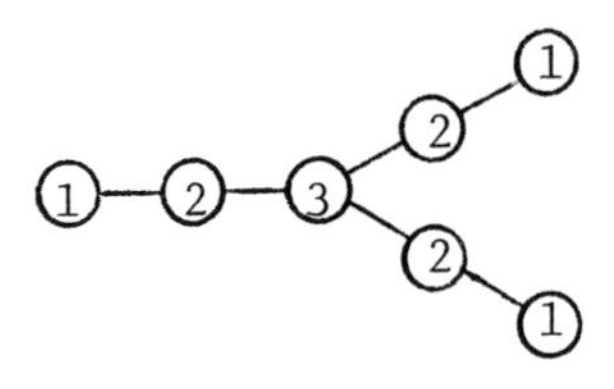

(XX.xi)

does the trick, where the μ value is indicated inside each circle.

For b.

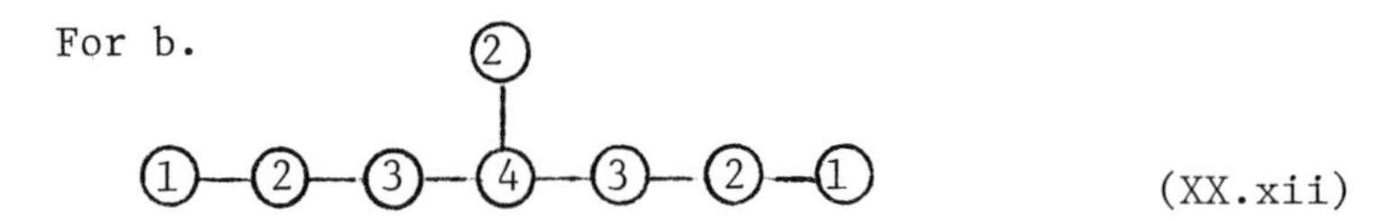

(XX.xii)

For c.

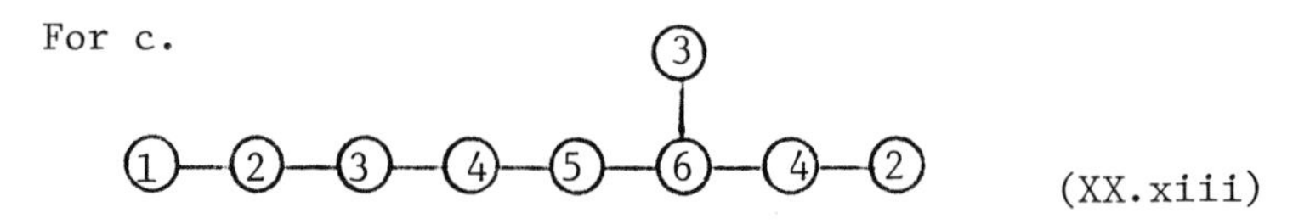

(XX.xiii)

In d, the vectors do not all have the same length. The two on the left are either longer or shorter than the others by a factor of $\sqrt{2}$. In the first case, use

(1)—(2)=(3)—(2)—(1) (XX.xiv)

for the second use

(2)—(4)=(3)—(2)—(1) (XX.xv)

Note (in passing) that all these diagrams can be constructed with linearly dependent vectors.

Putting all this together, we have learned that any Π-system must belong to one of 4 infinite families, or be one of 5 exceptional diagrams:

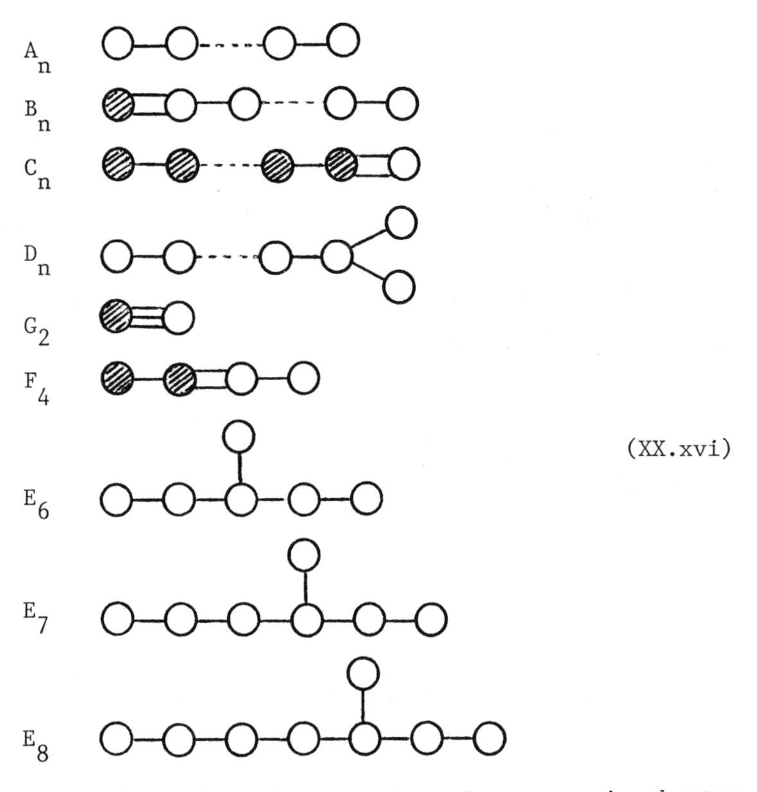

(XX.xvi)

Where the solid circle indicates that the vector is shorter than the others.

Obviously, the A_n are all allowed. Indeed, they are our old friends the unitary groups, $A_n = SU(n+1)$. With one branch, the D_n are allowed, but longer branches run afoul of (XX.xa-c) except for E_6, E_7, E_8. With one double line, the diagrams

(XX.xvii)

correspond to either B_n or C_n depending on whether the vector connected only to the double line is longer or shorter than the rest. Adding more vectors on the left conflicts with (XX.xd), except for F_4. And, of course, G_2 is the unique diagram with a triple line.

G_2

Let's find the roots corresponding to the Dynkin diagram

(XX.xviii)

Let α^1 and α^2 be the shorter and longer simple roots, respectively. They satisfy

$$\frac{2\alpha^1 \cdot \alpha^2}{\alpha^{2^2}} = -1, \quad \frac{2\alpha^1 \cdot \alpha^2}{\alpha^{1^2}} = -3. \qquad \text{(XX.4)}$$

For example, we can take

$$\alpha^1 = (0, \frac{1}{\sqrt{3}}), \quad \alpha^2 = (\frac{1}{2}, -\frac{\sqrt{3}}{2}) \qquad \text{(XX.5)}$$

(XX.xix) is shown on the next page. Now $\alpha^2+\alpha^1$, $\alpha^2+2\alpha^1$ and $\alpha^2+3\alpha^1$ are roots because

$$\frac{2\alpha^1 \cdot \alpha^2}{\alpha^{1^2}} = -3 = -p. \qquad \text{(XX.6)}$$

Calculate

$$\frac{2\alpha^2 \cdot (\alpha^2+3\alpha^1)}{\alpha^{2^2}} = -1 = -p, \qquad \text{(XX.7)}$$

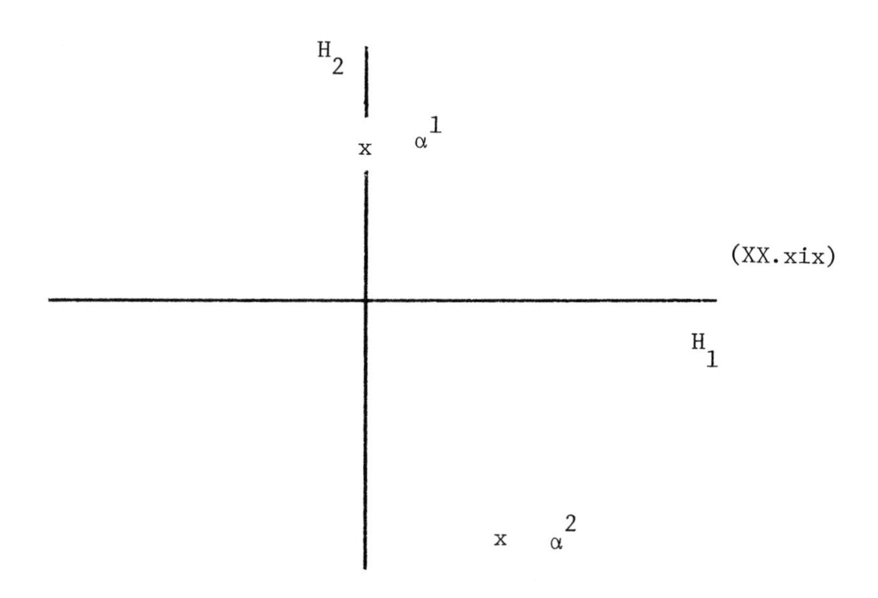

(XX.xix)

so $2\alpha^2+3\alpha^1$ is a weight (but not $3\alpha^2+3\alpha^1$). But

$$2\alpha^2 \cdot (\alpha^2+2\alpha^1) = 0 = -p \tag{XX.8}$$

so $2\alpha^2+2\alpha^1$ is not a weight. Now

$$\frac{2\alpha^2 \cdot (2\alpha^2+3\alpha^1)}{{\alpha^2}^2} = 1 = -p+1,$$

$$\frac{2\alpha^1 \cdot (2\alpha^2+3\alpha^1)}{{\alpha^1}^2} = 0 = -p. \tag{XX.9}$$

Thus, $2\alpha^2+3\alpha^1$ is the highest weight.

So, the positive roots are α^1, α^2, $\alpha^2+\alpha^1$, $\alpha^2+2\alpha^1$, $\alpha^2+3\alpha^1$ and $2\alpha^2+3\alpha^1$.

The root system is shown in (XX.xx) on the following page.

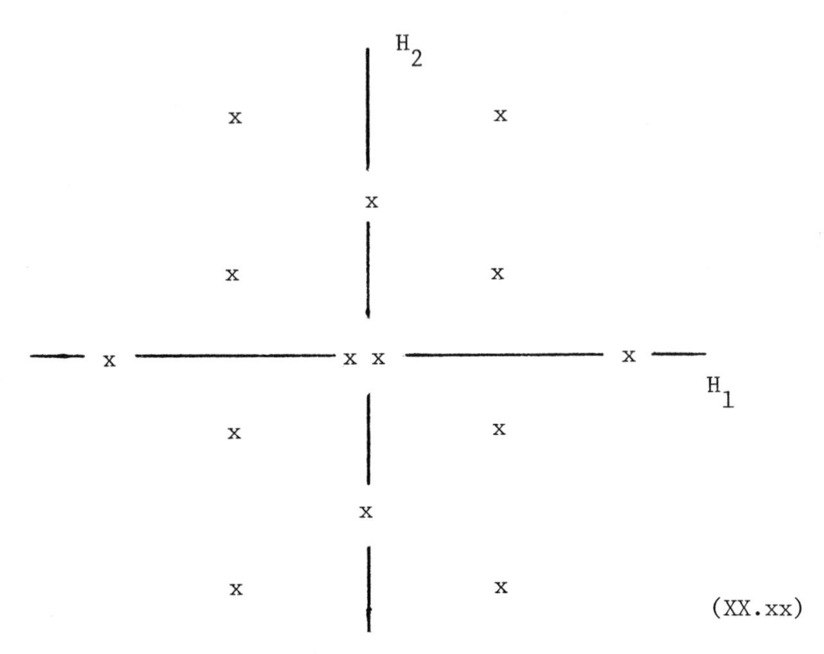

(XX.xx)

The highest weight of the adjoint representation, $2\alpha^2+3\alpha^1 = (1,\ 0) = \mu^2$ satisfies

$$\frac{2\alpha^2\cdot\mu^2}{{\alpha^2}^2} = 1, \quad \frac{2\alpha^1\cdot\mu^2}{{\alpha^1}^2} = 0 \tag{XX.10}$$

so it is a fundamental weight.

The other fundamental weight, μ^1, satisfies

$$\frac{2\alpha^2\cdot\mu^1}{{\alpha^2}^2} = 0, \quad \frac{2\alpha^1\cdot\mu^1}{{\alpha^1}^2} = 1. \tag{XX.11}$$

It is

$$\mu^1 = (\frac{1}{2}, \frac{1}{2\sqrt{3}}). \tag{XX.12}$$

Obviously, $\mu^1-\alpha^1$ is another weight in this representation. It satisfies

$$\frac{2\alpha^2 \cdot (\mu^1 - \alpha^1)}{{\alpha^2}^2} = 1 = q \qquad \text{(XX.13)}$$

hence $\mu^1 - \alpha^1 - \alpha^2$ is a weight, but not $\mu^1 - \alpha^1 - 2\alpha^2$.

$$\frac{2\alpha^1 \cdot (\mu^1 - \alpha^1 - \alpha^2)}{{\alpha^1}^2} = 2 = q \qquad \text{(XX.14)}$$

so $\mu - \alpha^1 - \alpha^2 - \alpha^1$ and $\mu - \alpha^1 - \alpha^2 - 2\alpha^1$ are weights but not $\mu - \alpha^1 - \alpha^2 - 3\alpha^1$. Using Weyl reflections, we find $\mu - 2\alpha^2 - 3\alpha^1$ and $\mu - 2\alpha^2 - 4\alpha^1$ are weights. All of these weights are singly degenerate. The states are obtained from the highest weight state $|\mu^1\rangle$ uniquely as follows:

$$E_{-\alpha^1}|\mu^1\rangle,\quad E_{-\alpha^2}E_{-\alpha^1}|\mu^1\rangle,\quad E_{-\alpha^1}E_{-\alpha^2}E_{-\alpha^1}|\mu^1\rangle,$$

$$(E_{-\alpha^1})^2\, E_{-\alpha^2}\, E_{-\alpha^1}|\mu^1\rangle,\quad E_{-\alpha^2}(E_{-\alpha^1})^2\, E_{-\alpha^2}\, E_{-\alpha^1}|\mu^1\rangle,$$

$$E_{-\alpha^1}\, E_{-\alpha^2}(E_{-\alpha^1})^2\, E_{-\alpha^2}\, E_{-\alpha^1}|\mu^1\rangle. \qquad \text{(XX.15)}$$

Thus, this representation is 7 dimensional.

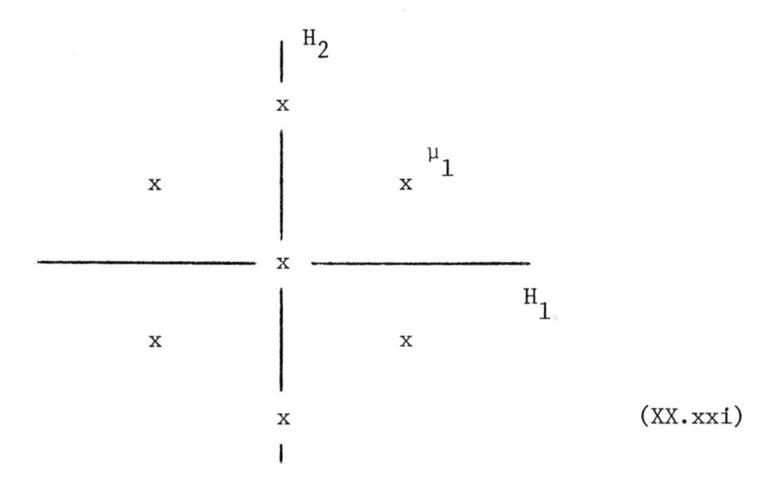

(XX.xxi)

Before the η meson was discovered, it was thought that G_2 might be a useful symmetry of strong interactions, with this 7 dimensional representation describing the π^+, π^0, π^-, K^+, K^0 and $\overline{K^0}$, K^- mesons. But, it didn't work out.

REGULAR SUBALGEBRAS

Notice that $\alpha^2 = (1/2, -\sqrt{3}/2)$ and $\alpha^2+3\alpha^1 = (1/2, \sqrt{3}/2)$ are the simple roots of SU(3). Indeed, E_{α^2}, $E_{\alpha^2+3\alpha^1}$, $E_{2\alpha^2+3\alpha^1}$, their complex conjugates and the Cartan subalgebra H_1 and H_2 generate an SU(3) subalgebra of G_2. Under this SU(3), the 14 dimensional adjoint representation of G_2 transforms like $8 \oplus 3 \oplus \bar{3}$, while the 7 dimensional representation is $3 \oplus \bar{3} \oplus 1$.

This SU(3) subalgebra is called "regular" and "maximal". A regular subalgebra, R, of a simple Lie algebra, A, is a subalgebra such that the roots of R are a subset of the roots of A and the generators of the Cartan subalgebra of R are linear combinations of the Cartan generators of A. A regular subalgebra is called maximal if the rank of R is the same as the rank of A (so that the Cartan subalgebras are the same).

Suppose we have the Dynkin diagram corresponding to some simple Lie algebras. We can find some regular subalgebras just by leaving out some of the simple roots. But the subalgebras obtained in this way are not maximal. To get the regular maximal subalgebras, consider the system of vectors formed by the simple roots α^i, i=1 to n and the lowest root (the negative of the highest root), call it α^0. Since $\alpha^0-\alpha^i$ is not a root $2\alpha^0\cdot\alpha^i/\alpha^{02}$ and $2\alpha^0\cdot\alpha^i/\alpha^{i2}$ are non-positive integers. Thus, the system of vectors α^k, $k = 0$ to n is like a Π-system except that there is one linear relation among the n+1 vectors. Such a system is called an extended Π-system. If we now remove any vector from the extended Π-system, the resulting vectors are the roots of a regular maximal subalgebra of the original algebra.

Given any Dynkin diagram, we can calculate the highest root and directly construct the extended Π-system. Clearly, there is a unique extended Π-system for each simple algebra. In fact, we have already met all the extended Π systems in the discussion of the classification theorem. Here we will give only the results on the next page.

The primes on the algebras indicate that these are the extended Dynkin diagrams. Note that the A_n's do not have any nontrivial regular maximal subalgebras. All the rest do. (XX.xxii) provides a very quick and useful way of finding these subgroups.

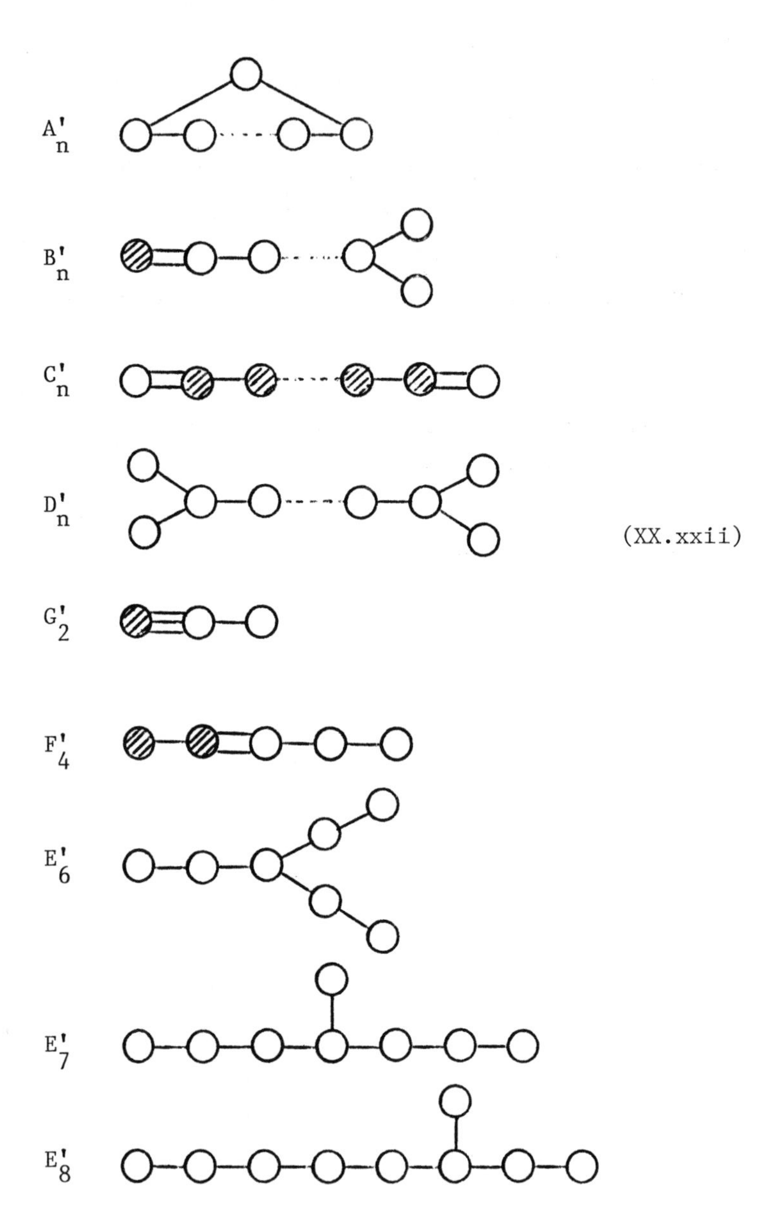

(XX.xxii)

PROBLEMS FOR CHAPTER XX

(XX.A) Prove that decomposable Π-systems yield decomposable root systems.

(XX.B) Find the regular maximal subalgebras of E_6.

(XX.C) Find the regular maximal subalgebras of SO(12). To find them all, you will have to apply the extended Dynkin diagram algorithm several times, because some of the regular maximal subalgebras themselves have nontrivial regular maximal subalgebras.

XXI. SO(2n+1) AND SPINORS

In the next five chapters, we are going to get to know the SO(N) algebras better. The peculiar thing about SO(N) is that it has spinor representations, generalizations of the Pauli matrices with similar properties. It is possible that these spinor representations play a role in physics. We will see in Chapter XXIV that the SU(5) of Chapter XVIII can be embedded in SO(10) in a very suggestive way, so that the creation operators for the particles

Howard Georgi, Lie Algebras in Particle Physics: From Isospin to Unified Theories ISBN 0-8053-3153-0

transform like the spinor representations.

We can label the generators of SO(2n+1) as follows

$$M_{ab} = -M_{ba} \quad \text{for} \quad a,\, b = 1 \quad \text{to} \quad 2n+1 \tag{XXI.1}$$

where in the 2n+1 dimensional defining representation discussed earlier, in Chapter XIX,

$$[M_{ab}]_{jk} = -i(\delta_{aj}\delta_{bk} - \delta_{bj}\delta_{ak}). \tag{XXI.2}$$

You can easily calculate the commutation relations:

$$[M_{ab},\, M_{cd}] = -i(\delta_{bc}M_{ad} - \delta_{ac}M_{bd} - \delta_{bd}M_{ac} + \delta_{ad}M_{bc}). \tag{XXI.3}$$

This is the standard form for rotation generators in a real vector space. The commuting generators in the Cartan sub-algebra, we took to be

$$H_j = M_{2j-1,2j},\; j = 1 \text{ to } n. \tag{XXI.4}$$

We can take the E generators to be

$$E_{\eta e^j} = \frac{1}{\sqrt{2}}\,(M_{2j-1,2n+1} + i\eta M_{2j,2n+1}),$$

$$E_{\eta e^j+\eta' e^k} = \frac{1}{2}\,(M_{2j-1,2k-1} + i\eta M_{2j,2k-1}) \tag{XXI.5}$$

$$+ i\eta' M_{2j-1,2k} - \eta\eta' M_{2j,2k}), \qquad \text{(XXI.5)}$$

where η, $\eta' = \pm 1$. These were chosen to satisfy

$$[E_{\eta e^j}, E_{\eta' e^k}] = i\, E_{\eta e^j + \eta' e^k}. \qquad \text{(XXI.6)}$$

We argued in Chapter XIX that the simple roots are

$$\alpha^i = e^i - e^{i+1} \quad \text{for} \quad i=1 \text{ to } n-1$$
$$\alpha^n = e^n. \qquad \text{(XXI.7)}$$

Corresponding to the diagram

$$\alpha^1 \quad \alpha^2 \quad \cdots \quad \alpha^n \qquad \text{(XXI.i)}$$

Look at the fundamental weights μ^1 and μ^n. Clearly,

$$\mu^1 = e^1. \qquad \text{(XXI.8)}$$

This is the highest weight of the defining representation, with weights $\pm e^j$ and 0. On the other hand,

$$\mu^n = (e^1 + e^2 + \cdots + e^n)/2. \qquad \text{(XXI.9)}$$

All the other weights in this fundamental representation can be obtained by Weyl reflections in the roots e^j. A general weight is

$$(\pm e^1 \pm e^2 \pm \cdots \pm e^n)/2. \qquad \text{(XXI.10)}$$

This representation is 2^n dimensional.

It is convenient to treat the 2^n dimensional space as a tensor product of n two dimensional spaces. Then an arbitrary matrix can be built up out of tensor products of Pauli matrices. Call the Pauli matrices for the jth space σ_a^j. In other words

$$|\pm \tfrac{1}{2} e^1 \pm \tfrac{1}{2} e^2 \pm \cdots \pm \tfrac{1}{2} e^n>.$$

$$\equiv |\pm \tfrac{1}{2} e^1> \otimes |\pm \tfrac{1}{2} e^2> \otimes \cdots \otimes |\pm \tfrac{1}{2} e^n>, \qquad \text{(XXI.11)}$$

$$\sigma_a^j |xe^j> = |x'e^j>[\sigma_a]_{x'x} \tag{XXI.11}$$

where x, x' = ± 1/2.

In this notation, the Cartan generators are

$$H_j = \frac{1}{2}\sigma_3^j. \tag{XXI.12}$$

Notice that these satisfy

$$H_j^2 = \frac{1}{4}. \tag{XXI.13}$$

Clearly, we could have chosen any M_{ab} to be a Cartan generator, so

$$(M_{ab})^2 = \frac{1}{4} \tag{XXI.14}$$

in this representation for any $a \neq b$. Consider

$$E_{e^j} = \frac{1}{\sqrt{2}}(M_{2j-1,2n+1} + i\, M_{2j,2n+1}). \tag{XXI.15}$$

Obviously $(E_{e^j})^2 = 0$, which implies

$$\left\{M_{2j-1,2n+1}, M_{2j,2n+1}\right\} = 0. \tag{XXI.16}$$

Again, the particular choice of axes, 2j-1, 2j and 2n+1 is purely conventional, so in this representation

$$\left\{M_{ik}, M_{jk}\right\} = 0, \quad i\neq j\neq k\neq i. \tag{XXI.17}$$

Now let us construct the E's. Clearly

$$E_{e^1}|-\frac{1}{2}e^1 + x_2e^2 + \cdots + x_ne^n>$$

$$= f(x_2\cdots x_n)|\frac{1}{2}e^1 + x_2e^2 + \cdots + x_ne^n>, \tag{XXI.18}$$

$$E_{-e^1}|-\frac{1}{2}e^1 + x_2e^2 + \cdots + x_ne^n> = 0. \tag{XXI.19}$$

For some coefficients $f(x_2\cdots x_n)$. Then

$$|f(x_2 \cdots x_n)|^2 = \langle -\tfrac{1}{2} e^1 + x_2 e^2 + \cdots + x_n e^n |$$

$$E_{-e^1} E_{e^1} | -\tfrac{1}{2} e^1 + x_2 e^2 + \cdots + x_n e^n \rangle$$

$$= \langle -\tfrac{1}{2} e^1 \cdots | \{E_{-e^1}, E_{e^1}\} | -\tfrac{1}{2} e^1 \cdots \rangle. \qquad \text{(XXI.20)}$$

But

$$\{E_{-e^1}, E_{e^1}\} = M^2_{1,2n+1} + M^2_{2,2n+1} = \frac{1}{2} \qquad \text{(XXI.21)}$$

so

$$|f(x_2 \cdots x_n)|^2 = \frac{1}{2}. \qquad \text{(XXI.22)}$$

For each $x_2 \cdots x_n$, we can choose the relative phase of the $\pm e^1/2$ states so that f is positive, so

$$f(x_2 \cdots x_n) = \frac{1}{\sqrt{2}} \qquad \text{(XXI.23)}$$

independent of x. Then

$$E_{\pm e^1} = \frac{1}{2\sqrt{2}} (\sigma^1_1 \pm i\sigma^1_2) = \frac{1}{2} \sigma^1_{\pm}. \qquad \text{(XXI.24)}$$

Now consider $E_{\pm e^2}$. By the same argument

$$E_{e^2} | x_1 e^1 - \tfrac{1}{2} e^2 + \cdots + x_n e^n \rangle$$

$$= f(x_1, x_3 \cdots x_n) | x_1 e^1 + \tfrac{1}{2} e^2 + \cdots \rangle. \qquad \text{(XXI.25)}$$

For $x_1 = 1/2$, we can choose the relative phases of the $\pm e^2/2$ states to make $f = 1/\sqrt{2}$.

$$E_{e^2} | \tfrac{1}{2} e^1 - \tfrac{1}{2} e^2 + \cdots \rangle$$

$$= \frac{1}{\sqrt{2}} | \tfrac{1}{2} e^1 + \tfrac{1}{2} e^2 + \cdots \rangle. \qquad \text{(XXI.26)}$$

But now the phases of the states with $x_1 = -1/2$ are fixed. In particular, since

$$\{E_{\pm e^1}, E_{\pm e^2}\} = 0,$$

$$E_{e^2}|-\frac{1}{2}e^1 - \frac{1}{2}e^2 + \cdots\rangle$$

$$= \sqrt{2}\, E_{e^2}E_{-e^1}|\frac{1}{2}e^1 - \frac{1}{2}e^2 + \cdots\rangle$$

$$= -\sqrt{2}\, E_{-e^1}E_{e^2}|\frac{1}{2}e^1 - \frac{1}{2}e^2 + \cdots\rangle$$

$$= -E_{-e^1}|\frac{1}{2}e^1 + \frac{1}{2}e^2 + \cdots\rangle$$

$$= -\frac{1}{\sqrt{2}}|-\frac{1}{2}e^1 + \frac{1}{2}e^2 + \cdots\rangle. \tag{XXI.27}$$

In other words,

$$E_{\pm e^2} = \frac{1}{2}\sigma_3^1\sigma_\pm^2, \tag{XXI.28}$$

(where the tensor product is understood).

Continuing in this way, we can write

$$E_{\pm e^3} = \frac{1}{2}\sigma_3^1\sigma_3^2\sigma_\pm^3 \tag{XXI.29}$$

$$\cdots$$

$$E_{\pm e^j} = \frac{1}{2}\sigma_3^1\cdots\sigma_3^{j-1}\sigma_\pm^j. \tag{XXI.30}$$

For the hermitian generators

$$M_{2j-1,2n+1} = \frac{1}{2}\sigma_3^1\cdots\sigma_3^{j-1}\sigma_1^j \tag{XXI.31}$$

$$M_{2j,2n+1} = \frac{1}{2}\sigma_3^1\cdots\sigma_3^{j-1}\sigma_2^j. \tag{XXI.32}$$

Now we know everything, since we can construct the other generators via

$$M_{ab} = -i[M_{a,2n+1}, M_{b,2n+1}], \tag{XXI.33}$$

for $a, b \neq 2n+1$. Each M_{ab} is just $\pm 1/2$ times a product of Pauli matrices.

This representation is called the spinor representation of SO(2n+1). It is the generalization of the spin 1/2 representation of SO(3) (SU(2)).

REAL AND PSEUDO-REAL

Suppose T_a generates a real irreducible representation of a simple Lie algebra, so that

$$T_a = -R\, T_a^* R^{-1}. \tag{XXI.34}$$

Let us first prove that R is unique up to trivial scale changes. Suppose there is another non-singular matrix Q, such that

$$T_a = -Q\, T_a^* Q^{-1} \tag{XXI.35}$$

then

$$T_a^* = -Q^{-1}\, T_a Q. \tag{XXI.36}$$

Thus,

$$T_a = R\, Q^{-1}\, T_a\, Q\, R^{-1} \tag{XXI.37}$$

which implies

$$[T_a,\, R\, Q^{-1}] = 0, \quad \text{for all a.} \tag{XXI.38}$$

But if a matrix commutes with all the generators of an irreducible representation, it must be a multiple of the unit matrix. This obvious fact is sometimes dignified by the title of Schur's Lemma. Let's quickly prove it.

Suppose $[T_a, M] = 0$. Write M in terms of hermitian matrices $M = h_1 + i\, h_2$. Then, $[T_a, h_i] = 0$, and we need only prove the result for a hermitian matrix, say h_1. Diagonalize h_1. Because $[T_a, h_1] = 0$, none of the generators connect states corresponding to different eigenvalues. But by irreducibility, all states are connected. So there is only one eigenvalue and $h = \lambda I$.

Returning to the problem at hand, we have shown

$$R\, Q^{-1} = \lambda I \tag{XXI.39}$$

thus $R = \lambda Q$ and R is essentially unique.

Because the T_a are hermitian, we can write

$$T_a = -R\, T_a^T\, R^{-1},\quad T_a^T = -R^{-1T}\, T_a R^T,$$

$$T_a = R\, R^{-1T}\, T_a R^T\, R^{-1},\quad [T_a, R^T R^{-1}] = 0. \tag{XXI.40}$$

Thus, $R^T R^{-1} = \lambda I$, or

$$R^T = \lambda R \tag{XXI.41}$$

but since $R = \lambda R^T$, we must have $\lambda = \pm 1$. So either

$$R = A = -A^T \quad \text{or} \quad R = S = S^T. \tag{XXI.42}$$

Thus, there are two kinds of real representations. If $R = S$ is symmetric, the representation is sometimes called real-positive or just plain real, while if $R = A$, it is called real-negative or pseudoreal.

To understand the difference, suppose T_a is equivalent to a representation by purely imaginary, antisymmetric matrices

$$T'_a = U\, T_a\, U^{-1},\quad T'_a = -T'^*_a. \tag{XXI.43}$$

Thus,

$$T'^T_a = U^{-1T}\, T_a^T\, U^T = -U\, T_a\, U^{-1},$$

$$T_a = -U^{-1} U^{-1T}\, T_a^T\, U^T U$$

$$= -(U^T U)^{-1}\, T_a^*(U^T U). \tag{XXI.44}$$

In this case, the matrix R is the symmetric matrix $U^T U$. Thus, a representation like the adjoint representation which is equivalent to a representation by pure imaginary matrices, is necessarily real-positive.

But, there are representations which are equivalent to their complex conjugates but which cannot be transformed into a representation by imaginary matrices. These are the pseudo-real representations. The simplest example is the spin 1/2 representation of SU(2) whose generators are the Pauli matrices, $T_a = \sigma_a/2$. Obviously, there is no way to generate this representation with antisymmetric imaginary 2x2 matrices, because there is only one such, while there are three generators. But, the representation is equivalent to its complex conjugate,

$$\frac{\sigma_a}{2} = -\sigma_2 \frac{\sigma_a^*}{2} \sigma_2, \tag{XXI.45}$$

and sure enough, the transformation matrix σ_2 is antisymmetric so the representation is pseudo-real.

There is another way of thinking about the matrix R. It is an invariant tensor. To see this, rewrite (XXI.34) as follows:

$$T_a R = -R T_a^*, \quad T_a R + R T_a^T = 0. \tag{XXI.46}$$

With explicit indices this becomes

$$(T_a)^i_j R^{jk} + (T_a)^k_j R^{ij} = 0. \tag{XXI.47}$$

Thus, R is an invariant tensor in the tensor product space of $D \otimes D$. For any real representation, therefore, $D \otimes D$ contains the trivial representation, 1, only once because R is unique. If D is real-positive, the coefficient of the representation 1 in the decomposition of $D \otimes D$ is symmetric in the exchange of the two equivalent D's (because it is proportional to R). If D is real-negative (pseudoreal), it is antisymmetric.

For the spinor representation of SO(2n+1), we can take the matrix R to be

$$\prod_{j\ \mathrm{odd}} \sigma_2^j \prod_{j\ \mathrm{even}} \sigma_1^j = R = R^{-1}. \qquad \text{(XXI.48)}$$

It is easy to check that

$$M_{j,2n+1} = -R\, M^*_{j,2n+1} R^{-1}. \qquad \text{(XXI.49)}$$

When the other generators are constructed as commutators of these, they obey the same relation.

Thus, for $n = 1$ or 2, R is antisymmetric; for $n = 3$ or 4, it is symmetric; for $n = 5$ or 6, it is antisymmetric again, etc. We can summarize this in the following table:

Algebra	Spinors	
SO(8k+3)	pseudo-real	
SO(8k+5)	pseudo-real	
SO(8k+7)	real	
SO(8k+1)	real	(XXI.ii)

PROBLEMS FOR CHAPTER XXI

(XXI.A) Show that the set of 10 matrices $\vec{\sigma}$, $\vec{\sigma}\tau_1$, $\vec{\sigma}\tau_3$ and τ_2 generate the spinor representation of SO(5). Find the matrix $R = R^{-1}$ such that $T_a = -R\, T_a^* R$ for this representation.

(XXI.B) Identify any convenient SO(2n−1) subalgebra of SO(2n+1) and determine how the spinor representation of SO(2n+1) transforms under the subalgebra.

XXII. SO(2n+2)

The standard notation for the even orthogonal groups is the same as (XXI.1-4) for the odd orthogonals, except that for SO(2n+2), j in (XXI.4) runs from 1 to n+1. We can use our results for SO(2n+1) to find the spinor representations of SO(2n+2).

The Dynkin diagram is

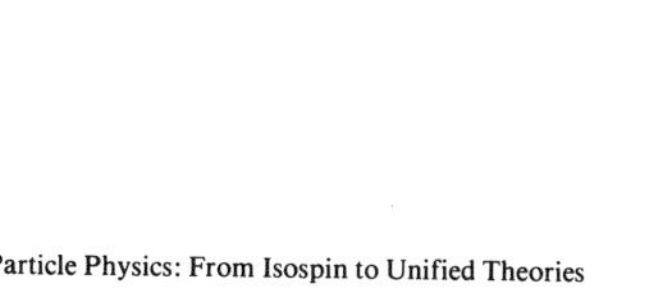

$\alpha^1 \quad \alpha^2 \quad \cdots \quad \alpha^n \quad \alpha^{n+1}$ (XXII.1)

Howard Georgi, Lie Algebras in Particle Physics: From Isospin to Unified Theories ISBN 0-8053-3153-0

where

$$\alpha^i = e^i - e^{i+1}, \quad i=1 \text{ to } n, \quad \alpha^{n+1} = e^n + e^{n+1}. \tag{XXII.1}$$

All the roots have the form

$$\pm e^i \pm e^j, \quad i \neq j. \tag{XXII.2}$$

We are interested in the two special representations corresponding to the last two fundamental weights.

$$\mu^n = (e^1 + e^2 + \cdots e^n - e^{n+1})/2,$$

$$\mu^{n+1} = (e^1 + e^2 + \cdots e^n + e^{n+1})/2. \tag{XXII.3}$$

Call the corresponding representations D^n and D^{n+1}. Clearly, the weights of D^n are of the form

$$\frac{1}{2} \sum_{i=1}^{n+1} \eta_i e^i \tag{XXII.4}$$

where

$$\eta_i = \pm 1 \quad \text{and} \quad \prod_{i=1}^{n+1} \eta_i = -1; \tag{XXII.5}$$

while for D^{n+1} they are given by (XXII.4) with

$$\prod_{i=1}^{n+1} \eta_i = 1. \tag{XXII.6}$$

Knowing the weights, we can discuss the reality properties of these representations. If n is odd, so that the algebra is of the form SO(4N), then $-\mu^n$ is a weight in D^n, because it satisfies (XXII.4-5). Clearly, it is the lowest weight in D^n. Similarly, $-\mu^{n+1}$ is the lowest weight in D^{n+1}. Thus, for n odd, the spinor representations of SO(2n+2) are real (or pseudoreal).

But for n even, $-\mu^n$ is the lowest weight in D^{n+1}, because it satisfies (XXII.6). And $-\mu^{n+1}$ is the lowest weight of D^n. Thus, for n even, the spinor representations are complex. D^n is $\bar{D}^{n+1}$ and D^{n+1} is $\bar{D}^n$.

To construct these representations in detail, consider the SO(2n+1) subgroup of SO(2n+2) generated by

$$M_{ij} \quad \text{for} \quad i, j \leq 2n+1. \tag{XXII.7}$$

This eliminates the Cartan generator

$$H_{n+1} = M_{2n+1,2n+2}. \tag{XXII.8}$$

The 4n generators with weights

$$\pm e^i \pm e^{n+1} \quad \text{for } i \leq n \tag{XXII.9}$$

which are linear combinations of

$$M_{j,2n+1} \text{ and } M_{j,2n+2} \quad j \leq 2n$$

collapse to the 2n generators $M_{j,2n+1}$ with weights

$$\pm e^i \quad \text{for } i \leq n. \tag{XXII.10}$$

Under this SO(2n+1) subgroup both D^n and D^{n+1} transform like the spinor representation of SO(2n+1) we analyzed in Chapter XXI. In fact, everything is the same as in the previous analysis, because H_{n+1} is not a generator of this subgroup so we just ignore the n+1 component of all weight vectors. Thus, the weight vectors for this subgroup are the same for both D^n and D^{n+1};

$$\frac{1}{2} \sum_{i=1}^{n} \eta_i e^i, \quad \eta_i = \pm 1. \tag{XXII.11}$$

In the tensor product notation, $\eta_i = \sigma_3^i$ and the generators are given by (XXI.31-33).

Now returning to the full SO(2n+2) group, note that in the representation D^n

$$\eta_{n+1} = -\prod_{i=1}^{n} \eta_i = -\sigma_3^1 \cdots \sigma_3^n. \tag{XXII.12}$$

Thus,

$$H_{n+1} = -\frac{1}{2}\sigma_3^1 \cdots \sigma_3^n = M_{2n+1,2n+2}. \tag{XXII.13}$$

Now all the SO(2n+2) generators can be found by commutation:

$$M_{j,2n+2} = i[M_{j,2n+1}, M_{2n+1,2n+2}]. \tag{XXII.14}$$

Similarly, in the representation D^{n+1},

$$\eta_{n+1} = +\prod_{i=1}^{n} \eta_i = \sigma_3^1 \cdots \sigma_3^n. \tag{XXII.15}$$

Thus,

$$H_{n+1} = +\frac{1}{2}\sigma_3^1 \cdots \sigma_3^n = M_{2n+1,2n+2}. \tag{XXII.16}$$

Now that we have explicitly constructed the representations, let us go back and exhibit the reality properties we discussed earlier. As with SO(2n+1), we define the matrix

$$R = R^{-1} = \prod_{j \text{ odd}} \sigma_2^j \prod_{j \text{ even}} \sigma_1^j. \tag{XXII.17}$$

Then examine the complex conjugates

$$-R\, T_a^* R^{-1} \tag{XXII.18}$$

for the representations D^n and D^{n+1}. We already know that

$$-R\, M^*_{ij} R^{-1} = M_{ij} \tag{XXII.19}$$

for i, j $\leq$ 2n+1.

For n odd,

$$-R\, M^*_{2n+1,2n+2} R^{-1} = M_{2n+1,2n+2} \tag{XXII.20}$$

and therefore (using the commutation relations)

$$-R\, T^*_a R^{-1} = T_a \tag{XXII.21}$$

for all the generators. Thus, as we already knew, these representations are real. For $n = 4k+1$, R is antisymmetric so D^n and D^{n+1} are pseudoreal. For $n = 4k+3$, R is symmetric and the representations are real-positive.

For n even,

$$-R\, M^*_{2n+1,2n+2}\, R^{-1} = -M_{2n+1,2n+2} \tag{XXII.22}$$

and the representations D^n and D^{n+1} get interchanged. So we can complete our table of spinor representations, (XXI.ii):

Algebra	Spinors
SO(8k+3)	pseudoreal
SO(8k+4)	pseudoreal
SO(8k+5)	pseudoreal
SO(8k+6)	complex
SO(8k+7)	real (positive)
SO(8k)	real (positive)
SO(8k+1)	real (positive)
SO(8k+2)	complex

(XXII.ii)

PROBLEMS FOR CHAPTER XXII

(XXII.A) Show that SO(2n) has a regular maximal subalgebra, SO(2m) x SO(2n-2m). How do the spinor representations of SO(2n) transform under the subgroup?

(XXII.B) Show that SO(2n+1) has a regular maximal subalgebra, SO(2m) x SO(2n-2m+1). How do the spinor representations of SO(2n+1) transform under the subgroup.

(XXII.C) Show that SO(4) has the same algebra as SU(2) x SU(2). Thus, it is not simple. Nevertheless, the arguments in this chapter apply. Explain how.

(XXII.D) Show that the SO(6) algebra and the SU(4) algebra are equivalent, with the 4 of SU(4) corresponding to a spinor representation of SO(6). In SU(4), $4 \otimes 4 = 6 \oplus 10$. The 6 is the vector representation of SO(6). What is the 10, in the SO(6) language?

XXIII. SU(n) $\subset$ SO(2n) AND CLIFFORD ALGEBRAS

Before discussing the embedding of SU(n) in SO(2n), we will discuss the spinor representations of SO(2n+1) and SO(2n) in a different language -- Clifford algebras. A Clifford algebra is a set of matrices satisfying the anti-commutation relations

$$\{\Gamma_j, \Gamma_k\} = 2\delta_{jk}, \quad j, k = 1 \text{ to } N. \tag{XXIII.1}$$

Howard Georgi, Lie Algebras in Particle Physics: From Isospin to Unified Theories ISBN 0-8053-3153-0

Given such a set of matrices, you can construct a representation of SO(N) by

$$M_{jk} = \frac{1}{4i}[\Gamma_j, \Gamma_k]. \tag{XXIII.2}$$

You can easily show that the M_{jk} have the SO(N) commutation relations and furthermore that

$$[M_{jk}, \Gamma_\ell] = i(\delta_{j\ell}\Gamma_k - \delta_{k\ell}\Gamma_j). \tag{XXIII.3}$$

Thus, the Γ's are a set of tensors transforming according to the N dimensional vector representation of SO(N).

For SO(2n+1), we can easily construct the Clifford algebra which yields the spinor representation we described. It is

$$\begin{aligned}
\Gamma_1 &= \sigma_2^1\sigma_3^2 \cdots \sigma_3^n \\
\Gamma_2 &= -\sigma_1^1\sigma_3^2 \cdots \sigma_3^n \\
\Gamma_3 &= \sigma_2^2\sigma_3^3 \cdots \sigma_3^n \\
\Gamma_4 &= -\sigma_1^2\sigma_3^3 \cdots \sigma_3^n \\
&\cdots\cdots \\
\Gamma_{2n-1} &= \sigma_2^n \\
\Gamma_{2n} &= -\sigma_1^n \\
\Gamma_{2n+1} &= \sigma_3^1\sigma_3^2 \cdots \sigma_3^n
\end{aligned} \tag{XXIII.4}$$

Note that the product

$$\Gamma_1 \Gamma_2 \cdots \Gamma_{2n+1} = i^n \qquad \text{(XXIII.5)}$$

is proportional to the identity.

For SO(2n), the spinor representations are 2^{n-1} dimensional, but it is impossible to build a Clifford algebra in a 2^{n-1} dimensional space. Only 2n-1 matrices can satisfy (XXIII.1). However, we can obviously construct a Clifford algebra in a 2^n dimensional space just by taking the first 2n of the Γ_j's in (XXIII.4).

Again, we can use this Clifford algebra to define SO(2n) generators. But now there is a nontrivial matrix which commutes with all the generators. It is

$$\Gamma_{2n+1} = (-i)^n \, \Gamma_1 \Gamma_2 \cdots \Gamma_{2n} = \prod_{j=1}^{n} \sigma_3^j . \qquad \text{(XXIII.6)}$$

Thus, as we suspected, this representation is not irreducible, but falls apart into two 2^{n-1} dimensional representations. For $\Gamma_{2n+1} = 1$ it is the spinor representation D^n. For $\Gamma_{2n+1} = -1$ it is D^{n-1}, the other spinor representation.

The Clifford construction tells us more than just how to build the spinor representation. It tells us about the Clebsch-Gordan decomposition of a product with two spinors.

First, let us discuss this decomposition in SO(2n+1). We can interpret (XXIII.3) as the statement that $(\Gamma_\ell)_{xy}$ is an invariant tensor. (XXIII.3) can be rewritten as

$$(M_{jk})_{xz}(\Gamma_\ell)_{zy} - (M^*_{jk})_{yz}(\Gamma_\ell)_{xz} + (M^{D^1}_{jk})_{\ell m}(\Gamma_m)_{xy} = 0, \qquad \text{(XXIII.7)}$$

where the

$$(M^{D^1}_{jk})_{\ell m} = -i(\delta_{j\ell}\delta_{km} - \delta_{jm}\delta_{k\ell}) \tag{XXIII.8}$$

generate the vector representation D^1 with highest weight μ^1. Thus, $(\Gamma_\ell)_{xy}$ is an invariant tensor in which the x index transforms like the spinor representation D^n, the y index like the conjugate representation $\bar{D}^n$ and the ℓ like the vector representation D^1. Of course, $\bar{D}^n$ is equivalent to D^n by a similarity transformation involving the matrix R, (XXI.48), but, we will continue to distinguish the two. We can always put the R's in at the end of the analysis if desired.

Equivalently, we can use $(\Gamma_\ell)_{xy}$ to pick out components of the tensor product $D^n \otimes \bar{D}^n$ which transform like a vector. Of course, the trivial representation also appears in $D^n \otimes \bar{D}^n$, through the invariant tensor δ_{xy}, the identity matrix.

What about the product of two Γ's. The symmetric product $\{\Gamma_j, \Gamma_k\}$ is proportional to the identity, so it is nothing new. The antisymmetric product is the commutator $[\Gamma_j, \Gamma_k]$, proportional to the generators themselves. Thus, as we already knew, the adjoint representation is in $D^n \times \bar{D}^n$. But, the adjoint representation is the antisymmetric tensor representation with two indices. This is not hard to see directly. Just use the Jacobi identity,

$$[M_{jk}, [\Gamma_\ell, \Gamma_m]] = [[M_{jk}, \Gamma_\ell], \Gamma_m] + [\Gamma_\ell, [M_{jk}, \Gamma_m]], \tag{XXIII.9}$$

which clearly shows that the antisymmetric product of Γ's transforms like an antisymmetric product of vectors.

In general, consider the product of $m \leq n$ Γ's. We can ignore everything but the completely antisymmetric product

$$[\Gamma_{j_1} \cdots \Gamma_{j_m}] \equiv \sum_{P\begin{pmatrix} k_1 \cdots k_m \\ j_1 \cdots j_m \end{pmatrix}} \pm \Gamma_{k_1} \cdots \Gamma_{k_m} \qquad \text{(XXIII.10)}$$

with the - sign for odd permutations. Everything else can be reduced (using the anticommutation relations) to a product of m-2 or fewer Γ's. Again, the Jacobi identity can be used to show that this antisymmetric product of Γ's transforms like an antisymmetric product of vectors--an antisymmetric tensor with m indices.

We do not get any new matrices for $m > n$, because of the fact that the product of all 2n+1 Γ's is proportional to the identity.

Thus, the tensor product of two spinors in SO(2n+1) can be decomposed into antisymmetric tensors of rank 0 (the trivial representation), 1 (the vector), 2 (the adjoint for $n > 1$), $\cdots$ to n. You can check that the dimensions work by using the binomial theorem. Note, also, that this explains why the rank n antisymmetric tensor is not fundamental. It is the representation of highest weight in the tensor product of two spinors (can you see this directly from the form of the fundamental weights, (XXI.9)?).

For SO(2n) the analysis is somewhat more complicated because the 2^n dimension space on which the Clifford algebra is defined is not an irreducible representation. Let us define projection operators onto the irreducible subspaces.

$$P_{\pm} = \frac{1}{2}(1 \pm \Gamma_{2n+1}). \qquad \text{(XXIII.11)}$$

P_+ projects onto D^n while P_- projects onto D^{n-1}. Now a general $2^n \times 2^n$ matrix transforms under commutation with the generators like the tensor product

$$(D^n \oplus D^{n-1}) \otimes (\bar{D}^n \oplus \bar{D}^{n-1}) \qquad \text{(XXIII.12)}$$

and we can project out all four possibilities with the $P^{\pm}$. If K is an arbitrary matrix:

$$\begin{aligned} &P^+ K P^+ \quad \text{transforms like} \quad D^n \otimes \bar{D}^n; \\ &P^- K P^- \quad \text{transforms like} \quad D^{n-1} \otimes \bar{D}^{n-1}; \\ &P^+ K P^- \quad \text{transforms like} \quad D^n \otimes \bar{D}^{n-1}; \\ &P^- K P^+ \quad \text{transforms like} \quad D^{n-1} \otimes \bar{D}^n. \end{aligned} \qquad \text{(XXIII.13)}$$

As before, we can construct the K matrices out of antisymmetric products of the Γ's. We need not include Γ_{2n+1}, because the projection operators make it ± 1. Furthermore, only the even or odd products contribute in each projection, even in $D^n \otimes \bar{D}^n$ and $D^{n-1} \otimes \bar{D}^{n-1}$ and odd in the others. Again, we can ignore tensors of rank $m > n$.

The only other subtlety occurs for $m = n$ where not all components of the nth rank antisymmetric tensor in $D^n \otimes \bar{D}^n$ or $D^{n-1} \otimes \bar{D}^{n-1}$ are independent.

For example, in $D^n \otimes \bar{D}^n$

$$\begin{aligned} P^+ \Gamma_1 \Gamma_2 \cdots \Gamma_n P^+ &= P^+ \Gamma_{2n+1} \Gamma_1 \Gamma_2 \cdots \Gamma_n P^+ \\ &= P^+ (-i)^n \Gamma_1 \cdots \Gamma_{2n} \Gamma_1 \cdots \Gamma_n P^+ \\ &= (i)^n (-1)^{n(n-1)/2} P^+ \Gamma_{n+1} \Gamma_{n+2} \cdots \Gamma_{2n} P^+ \\ &= (i)^n (-1)^{n(n-1)/2} \frac{1}{n!} \varepsilon_{12\cdots n j_1 \cdots j_n} \end{aligned} \qquad \text{(XXIII.14)}$$

$$P^+\Gamma_{j_1}\cdots\Gamma_{j_n}P^+. \qquad \text{(XXIII.14)}$$

Clearly, the rank n tensors satisfy a self-duality condition, the nature of which depends on n. If n is even, the relation is real, so the representation is real. If n is odd, however, the relation is complex so the representation is complex. It is through this complex self-duality condition that the complexity of the SO(4n+2) representations enters the tensors (see problem (XXIII.C)).

This is summarized below, incorporating the fact that $\bar{D}^n = D^n$ and $\bar{D}^{n-1} = D^{n-1}$ for SO(2n) with n even while $\bar{D}^n = D^{n-1}$ for n odd. We denote the rank m antisymmetric tensor by (m)

$$SO(2n+1):\quad D^n \otimes D^n = \sum_{m=0}^{n} (m). \qquad \text{(XXIII.15)}$$

$$SO(4n):\quad D^n \otimes D^n = \sum_{m=0}^{n-1} (2m) + (2n)_1;$$

$$D^{n-1} \otimes D^{n-1} = \sum_{m=0}^{n-1} (2m) + (2n)_2;$$

$$D^n \otimes D^{n-1} = \sum_{m=0}^{n-1} (2m+1); \qquad \text{(XXIII.16)}$$

where $(2n)_{1,2}$ are real, self-dual and anti-self-dual.

$$SO(4n+2):\quad D^n \otimes D^n = \sum_{m=0}^{n-1} (2m+1) + (2n+1)_1;$$

$$D^{n-1} \otimes D^{n-1} = \sum_{m=0}^{n-1} (2m+1) + (2n+1)_2;$$

$$D^n \otimes D^{n-1} = \sum_{m=0}^{n} (2m); \qquad \text{(XXIII.17)}$$

where $(2n+1)_{1,2}$ satisfy complex self-duality conditions.

Now, we can identify an SU(n) subgroup of SO(2n). First, construct the objects

$$A_j = (\Gamma_{2j-1} - i\,\Gamma_{2j})/2. \tag{XXIII.18}$$

Because of the Clifford algebra, these satisfy

$$\{A_j, A_k\} = 0 = \{A_j^\dagger, A_k^\dagger\}, \ \{A_j, A_k^\dagger\} = \delta_{jk}. \tag{XXIII.19}$$

They are a set of creation and annihilation operators! We know how to form some SU(n) generators from such things

$$T^a = \sum_{j,k} A_j^\dagger\, [T^a]_{jk}\, A_k \tag{XXIII.20}$$

where $[T^a]_{jk}$ are the traceless hermitian nxn matrices.

We know that T^a generate SU(n) on the 2^n dimensional space of the Clifford Algebra, but we have not shown that it is a subalgebra of SO(2n). But note that

$$A_j^\dagger A_k = \frac{1}{2}\,\delta_{jk} + \frac{i}{2}\,M_{2j-1,2k-1} + \frac{1}{2}\,M_{2j-1,2k}$$
$$- \frac{1}{2}\,M_{2j,2k-1} + \frac{i}{2}\,M_{2j,2k}. \tag{XXIII.21}$$

The δ_{jk} term does not contribute to T_a because $[T_a]_{jk}$ is traceless. The other terms are SO(2n) generators. So, the T_a are a subset of the SO(2n) generator. They therefore generate a subalgebra.

The A_j, being annihilation operators, annihilate something. In this space it is the state with $\sigma_3^j = -1$ for all j. Call this the state $|0\rangle$. It is in the representation D^n for n even and D^{n-1} for n odd.

Note, also, that the creation operators $A_j^\dagger$ are tensor operators:

$$[T^a, A_j^\dagger] = \sum_k A_k^\dagger [T^a]_{kj}. \tag{XXIII.22}$$

They transform according to the n dimensional defining representation of SU(n).

When we act on $|0\rangle$ with some number of $A^\dagger$'s (say m), we therefore get a set of states transforming like an antisymmetric tensor product of m n's. This is the representation [m]. The representations are antisymmetric because the $A^\dagger$'s anticommute. Otherwise, this construction is similar to the n dimensional harmonic oscillator discussed in Chapter XIV.

Note, also, that if m is even, the states formed by acting on $|0\rangle$ with m $A^\dagger$'s are in the same irreducible representation of SO(2n) as the $|0\rangle$ state, because Γ_{2n+1} anticommutes with each $A^\dagger$, so it commutes with the product of m of them. But for m odd, the states are in the other irreducible SO(2n) representation.

Putting all this together, we can summarize the embedding of SU(n) in SO(2n) as follows

$$SO(4n+2): \quad D^{2n+1} = \sum_{j=0}^{n} [2j+1], \quad D^{2n} = \sum_{j=0}^{n} [2j]. \tag{XXIII.23}$$

Note that $\bar{D}^{2n} = D^{2n+1}$ because $[\bar{j}] = [2n+1-j]$

$$SO(4n): \quad D^{2n} = \sum_{j=0}^{n} [2j], \quad D^{2n-1} = \sum_{j=0}^{n-1} [2j+1]. \tag{XXIII.24}$$

There is an SO(2n) generator which commutes with all of the generators of the SU(n) subgroup we constructed above. It is

$$S = \sum_{j=1}^{n} M_{2j-1,2j} = \sum_{j=1}^{n} A_j^\dagger A_j - \frac{n}{2}. \tag{XXIII.25}$$

In the space of the Clifford Algebra, this is simply

$$S = \frac{1}{2} \sum_{j=1}^{n} \sigma_3^j. \tag{XXIII.26}$$

Thus $S|0\rangle = -n/2\ |0\rangle$. Furthermore, $[S, A^{\dagger}_j] = A^{\dagger}_j$, thus the creation operators raise S.

So the states in the representation [m] in the spinor representation of SO(2n) have $S = m-n/2$.

PROBLEMS FOR CHAPTER XXIII

(XXIII.A) Check that the dimensions work out in (XXIII.15 and 16) by using the binomial theorem.

(XXIII.B) Use (XXI.2) and (XXIII.2, 20 and 21) to determine how the vector representation D^1 of SO(2n) transforms under SU(n). Explain why this result is obvious.

(XXIII.C) Let $u^{jk\ell}$ be a completely antisymmetric tensor in SO(6). A self-duality condition has the form

$$u^{jk\ell} = \lambda\, \varepsilon^{jk\ell abc} u^{abc} .$$

What are the possible values of λ?

XXIV. SO(10)

Consider (XXIII.23) for $n = 2$, or SO(10). It shows that the SU(5) content of the spinor representations is as follows:

$$D^5 = [1] + [3] + [5] = 5 + \overline{10} + 1,$$

$$D^4 = [0] + [2] + [4] = 1 + 10 + \bar{5}. \qquad \text{(XXIV.1)}$$

Howard Georgi, Lie Algebras in Particle Physics: From Isospin to Unified Theories ISBN 0-8053-3153-0

Comparing (XXIV.1), in turn, with (XVIII.15 and 18), we see that D^5 has the right SU(5) content to describe the right-handed creation operators plus one additional SU(5) singlet, and D^4 behaves like the left-handed creation operators plus a singlet.

Thus, SO(10) may represent another stage in the unification process. It incorporates all the creation operators in a single irreducible representation. Of course, it also incorporates an extra singlet which does not seem to correspond to any of the particles we discussed in Chapter XX. In fact, however, the extra singlet is related to the particles which are actually observed in an interesting way. When we include it, we can restore the parity, or mirror symmetry that is lost in SU(5). To see what is happening, let us examine another subgroup of SO(10).

The SU(5) subgroup of SO(10) is regular but not maximal. It is obtained by removing one of the circles from the Dynkin diagram. To find the regular maximal subgroups, consider the extended Dynkin diagram, from (XX.xvi and xxii).

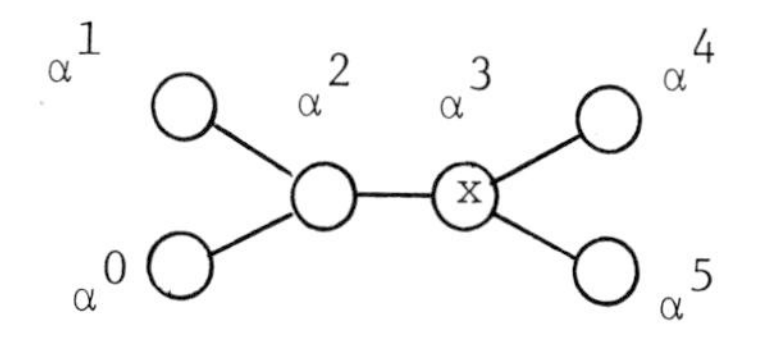

(XXIV.i)

If we remove the circle labeled with an x, the diagram falls apart into SU(2) x SU(2) x SU(4). The SU(2) x SU(2) is the same algebra as SO(4) and the SU(4) is the same as SO(6), thus this is the subalgebra of the 10 dimensional rotation generators which are block diagonal on 4 and 6 dimensional subspaces.

We can use the Dynkin diagrams rather directly to see how the spinor representation transforms under the subgroup. The weights of D^5 are

$$(\eta_i e^i)/2, \quad \Pi \eta_i = 1 \tag{XXIV.2}$$

from (XXII.4-6). The SU(4) (or SO(6)) subgroup has roots, α^1, α^2 and α^0 (the lowest root) which are

$$e^1-e^2, \; e^2-e^3 \quad \text{and} \quad -e^1-e^2. \tag{XXIV.3}$$

Thus, it acts on the first three components of the weight space. The weights, (XXIV.2) decompose into two copies of each of the two spinor representations of SO(6), one for $\eta_1\eta_2\eta_3 = 1$, the other $\eta_1\eta_2\eta_3 = -1$.

The SU(2) x SU(2) (or SO(4)) subgroup corresponds to the roots

$$\alpha^4 = e^4-e^5, \; \alpha^5 = e^4+e^5. \tag{XXIV.4}$$

These act on the last two components of the weight space. α^4 acts on weights of the form

$$\pm(e^4-e^5)/2. \tag{XXIV.5}$$

α^5 acts on weights of the form

$$\pm(e^4+e^5)/2. \tag{XXIV.6}$$

Thus, the weights

$$\sum_{i=1}^{3} (\eta_i e^i)/2 \pm (e^4+e^5)/2, \tag{XXIV.7}$$

are associated with a representation of the subalgebra which transforms like a spinor with respect to the SO(6) (or the 4, say, of SU(4)), like a singlet under the SU(2) generated by $E_{\pm\alpha^4}$ and like a doublet under the SU(2) generated by $E_{\pm\alpha^5}$, which we will call SU(2)'. Thus, under the SU(4) x SU(2) x SU(2)' subgroup, (XXIV.7) is a (4, 1, 2).

The weights

$$\sum_{i=1}^{3} (\eta_i e^i)/2 \pm (e^4 - e^5)/2, \quad \Pi \eta_i = -1, \tag{XXIV.8}$$

are associated with a representation of the subalgebra which transforms like the complex conjugate of SO(6) (or the $\bar{4}$ of SU(4)), like a doublet under SU(2) and a singlet under SU(2)'. Thus, it is a ($\bar{4}$, 2, 1). Hence,

$$D^n \to (4, 1, 2) \oplus (\bar{4}, 2, 1) \tag{XXIV.9}$$

under SU(4) x SU(2) x SU(2)'.

This SU(4) contains color SU(3). Each 4 is a $3 \oplus 1$ under SU(3). Each $\bar{4}$ is a $\bar{3} \oplus 1$. Comparing (XXIV.9) with (XVIII.9), you can see in more detail how the SO(10) unification works. The weak interaction SU(2) (XVIII.9) must be identified with the SU(2) subgroup of SO(10). Under it, the creation operators for the right-handed antiquarks and the right-handed positron and antineutrino transform as doublets. Under the SU(2)' of SO(10), the creation operators for the right-handed quarks are a doublet, and there is another doublet comprising the creation operators for the right-handed electron and a neutral particle. This SU(2)' is a mirror image of the weak interaction SU(2). If we look at the creation operators for the left-handed fields, the antiquarks will be a doublet under SU(2)', just as the right-handed antiquarks are a doublet under SU(2).

Thus, we should identify the neutral partner of the right-handed electron in the SU(2)' doublet as a right-handed neutrino, the looking glass version of the left-handed neutrino.

The SO(10) unification thus restores the mirror symmetry of the theory that was lost completely in SU(5). Of course, physics is not mirror symmetric (parity is violated). The right-handed neutrino has never been seen, nor is there any other evidence for the SU(2)' symmetry. Presumably, if the SO(10) unification is really relevant, there is more spontaneous symmetry breaking such that the right-handed neutrino and the particles associated with the SU(2)' generators (and all the other SO(10) generators except those of the SU(3) x SU(2) x U(1) subgroup) are very heavy.

Or maybe Nature does not care about irreducible representations and mirror symmetry, and SU(5) is the whole story of the particle interactions. We simply do not know yet.

PROBLEMS FOR CHAPTER XXIV

(XXIV.A) Show that the matrices $\vec{\sigma}$, $\vec{\tau}$, $\vec{\eta}$, $\vec{\sigma}\rho_1$, $\vec{\tau}\rho_2$, $\vec{\eta}\rho_3$, $\vec{\sigma}\vec{\tau}\rho_3$, $\vec{\tau}\vec{\eta}\rho_1$ and $\vec{\eta}\vec{\sigma}\rho_2$ where σ, τ, η and ρ are independent Pauli matrices, generate a spinor representation of SO(10). Find an SU(2) x SU(2) x SU(4) subgroup in which one of the SU(2) factors is generated by the subset $\vec{\eta}(1+\rho_3)/4$.

(XXIV.B) What is the dimension of the SO(10) representation with highest weight $2\mu^5$. How do you know? Hint: consider $D^5 \otimes D^5$.

XXV. AUTOMORPHISMS

An <u>automorphism</u> A of a group of G is a mapping of the group onto itself which preserves the group multiplication rule:

$$A(g_1 g_2) = A(g_1) \cdot A(g_2). \tag{XXV.1}$$

For a Lie group, an automorphism of the group induces a

Howard Georgi, Lie Algebras in Particle Physics: From Isospin to Unified Theories ISBN 0-8053-3153-0

mapping of the Lie algebra onto itself which preserves the commutation relations. Under an automorphism, the generators are mapped into linear combinations of generators

$$T^a \to A^{ab} T^b \tag{XXV.2}$$

such that $[T^a, T^b] = i\, f^{abc} T^c$ implies

$$[A^{aa'} T^{a'}, A^{bb'} T^{b'}] = i\, f^{abc} A^{cc'} T^{c'} . \tag{XXV.3}$$

Some automorphisms are trivial in the sense that the mapping they induce on the generators is an equivalence:

$$A^{ab} T^b = R\, T^a R^{-1} . \tag{XXV.4}$$

But some of the Lie algebras have non-trivial automorphisms.

For example, consider complex conjugation. If T^a are the generators of some representation, the mapping

$$T^a \to -T^{a*} \tag{XXV.5}$$

is an automorphism in which generators corresponding to imaginary antisymmetric matrices are unchanged while generators corresponding to real symmetric matrices change sign (note that we can always choose the generators to be either antisymmetric or symmetric in the highest weight construction). Thus, an algebra can have complex representations only if it has some nontrivial automorphism.

We can identify the nontrivial automorphisms by looking at the symmetries of the Dynkin diagram. For example, consider SU(4) whose Dynkin diagram is

$$\alpha^1 \quad \alpha^2 \quad \alpha^3 \qquad \text{(XXV.i)}$$

Since α^1 and α^3 appear symmetrically, the diagram with α^1 and α^3 interchanged has the same angles and therefore generates the same algebra. So, there is an automorphism of SU(4) in which the corresponding generators are interchanged

$$E_{\alpha^1} \leftrightarrow E_{\alpha^3}, \qquad \text{(XXV.6)}$$

along with all the other changes this induces in the explicit construction of the algebra from the diagram. This automorphism is clearly nontrivial, since it interchanges the fundamental representations D^1 and D^3. In fact, since D^1 is the 4 dimensional defining representation and D^3 is its complex conjugate representation $\bar{4}$, this is clearly just the automorphism induced by complex conjugation of the 4, up to some trivial equivalence. All of the complex conjugation automorphisms are obtained in this way.

As another example, consider one of the orthogonal groups we just discussed, SO(10), whose Dynkin diagram is

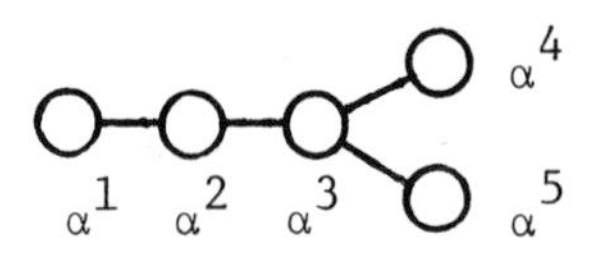

(XXV.ii)

There is an obvious symmetry under the exchange $\alpha^4 \leftrightarrow \alpha^5$, and sure enough, we saw that the fundamental (16 dimensional) representations D^4 and D^5 are complex conjugates of one another.

There are also nontrivial automorphisms which do not correspond to complex conjugation, as in the groups SO(4n), where the Dynkin diagrams are symmetric in the exchange $\alpha^{2n-1} \leftrightarrow \alpha^{2n}$, but the corresponding representations are both real.

Finally, there is the bizarre case of SO(8), with Dynkin diagram

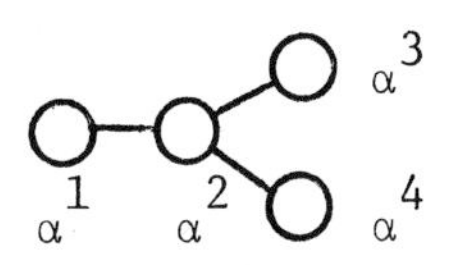

(XXV.iii)

Here, there are automorphisms for each permutation of α^1, α^3 and α^4. Let us discuss this one in more detail.

FUN WITH SO(8)

The roots in the Dynkin diagram (XXV.iii) are

$$\alpha^i = e^i - e^{i+1} \quad \text{for } i = 1 \text{ to } 3, \; \alpha^4 = e^3 + e^4 . \tag{XXV.7}$$

The fundamental representations D^1, D^3 and D^4 with highest weights μ^1, μ^3 and μ^4 are 8 dimensional, D^1 being the defining representation (an 8-vector), D^3 and D^4 are the real spinor representations. We want to analyze the automorphisms which interchange these various representations. We will first discuss the connection between D^3 and D^4.

Obviously, we want to interchange α^3 and α^4, with α^1 and α^2 fixed, which is implemented by changing the sign of e^4. That means

$$H_4 \to -H_4 \quad \text{or} \quad M_{78} \to -M_{78} . \tag{XXV.8}$$

Thus,

$$E_{\eta e^i+e^4} \leftrightarrow E_{\eta e^i-e^4}. \qquad \text{(XXV.9)}$$

But this means

$$M_{i8} \to -M_{i8}. \qquad \text{(XXV.10)}$$

Indeed, this is what we found when we explicitly constructed these representations. M_{ij} for i, j = 1 to 7 were the same for both representations, while

$$H_4 = M_{78} = -\frac{1}{2}\sigma_3^1\sigma_3^2\sigma_3^3 \text{ for } D^3, = \frac{1}{2}\sigma_3^1\sigma_3^2\sigma_3^3 \text{ for } D^4, \qquad \text{(XXV.11)}$$

(see (XXII.13 and 18). The rest follows from commutation. Note that we got the result for the automorphism directly from the Dynkin diagram, without explicitly constructing the representation.

Now, let's examine the automorphism which interchanges the spinor representation D^3 with the vector representation D^1. This is bound to be a rather peculiar mapping, since the generators of the spinor representation are products of Pauli matrices satisfying $M_{ij}^2 = 1/4$, while the generators of the vector representation D^1 are the imaginary antisymmetric matrices. The relevant mapping as identified from (XXV.iv) is $\alpha^1 \leftrightarrow \alpha^3$, α^2, α^4 fixed. In terms of the basis vectors the mapping is

$$e^1 - e^2 \leftrightarrow e^3 - e^4,\ e^2 - e^3 \to e^2 - e^3,\ e^3 + e^4 \to e^3 + e^4. \qquad \text{(XXV.12)}$$

Or,

$$e^1 \to \frac{1}{2}(e^1 + e^2 + e^3 - e^4),$$

$$e^2 \to \frac{1}{2}(e^1 + e^2 - e^3 + e^4),$$

$$e^3 \to \frac{1}{2}(e^1 - e^2 + e^3 + e^4),\quad e^4 \to \frac{1}{2}(-e^1 + e^2 + e^3 + e^4) \qquad \text{(XXV.13)}$$

So the Cartan generators are mapped as follows

$$H_1 \to \frac{1}{2}(H_1 + H_2 + H_3 - H_4),$$
$$H_2 \to \frac{1}{2}(H_1 + H_2 - H_3 + H_4),$$
$$H_3 \to \frac{1}{2}(H_1 - H_2 + H_3 + H_4),$$
$$H_4 \to \frac{1}{2}(-H_1 + H_2 + H_3 + H_4); \qquad \text{(XXV.14)}$$

or

$$M_{12} \to \frac{1}{2}(M_{12} + M_{34} + M_{56} - M_{78}),$$
$$M_{34} \to \frac{1}{2}(M_{12} + M_{34} - M_{56} + M_{78}),$$
$$M_{56} \to \frac{1}{2}(M_{12} - M_{34} + M_{56} + M_{78}),$$
$$M_{78} \to \frac{1}{2}(-M_{12} + M_{34} + M_{56} + M_{78}). \qquad \text{(XXV.15)}$$

Notice how the theory has solved the problem we mentioned earlier. The antisymmetric matrices $M_{2i-1,2i}$ of the vector representation are mapped into matrices of the form

$$\frac{1}{2}\begin{pmatrix} \pm\sigma_2 & & & 0 \\ & \pm\sigma_2 & & \\ & & \pm\sigma_2 & \\ 0 & & & \pm\sigma_2 \end{pmatrix} \qquad \text{(XXV.16)}$$

whose square is 1/4.

It should be clear how the rest of the mappings go. The 28 generators break up into 7 sets of 4, each of which mix up among themselves like the four M_{12}, M_{34}, M_{56} and M_{78} above. Consider, for example, that the mapping $e^1 - e^2 \leftrightarrow e^3 - e^4$, $e^1 + e^2$ and $e^3 + e^4$ unchanged, implies $E_{\pm e^1 \mp e^2} \leftrightarrow E_{\pm e^3 \mp e^4}$, $E_{\pm e^1 \pm e^2}$ and $E_{\pm e^3 \pm e^4}$ unchanged. So,

$$\{M_{13} \pm M_{23} \mp i\,M_{14} + M_{24}\}$$

$$\leftrightarrow \{M_{57} \pm i\,M_{67} \mp i\,M_{58} + M_{68}\}, \qquad \text{(XXV.17)}$$

while

$$\{M_{13} \pm i\,M_{23} \pm i\,M_{14} - M_{24}\},$$

$$\{M_{57} \pm i\,M_{67} \pm i\,M_{58} - M_{68}\} \qquad \text{(XXV.18)}$$

remain unchanged. Thus,

$$M_{13} + M_{24} \leftrightarrow M_{57} + M_{68}, \qquad \text{(XXV.18)}$$

$M_{13} - M_{24}$ and $M_{57} - M_{68}$ are unchanged, and

$$M_{23} - M_{14} \leftrightarrow M_{67} - M_{58}, \qquad \text{(XXV.20)}$$

$M_{23} + M_{14}$ and $M_{67} + M_{58}$ are unchanged. Thus, M_{13}, M_{24}, M_{57} and M_{68} mix up among themselves. For example, $M_{13} \to \frac{1}{2}(M_{13} - M_{24} + M_{57} + M_{68})$. Similarly, M_{23}, M_{14}, M_{67} and M_{58} mix.

From the mappings $e^1 - e^3 \leftrightarrow e^2 - e^4$, $e^1 + e^3$ and $e^2 + e^4$ unchanged, we obtain the sets M_{15}, M_{26}, M_{37} and M_{48}; and M_{25}, M_{16}, M_{47} and M_{38} which mix. While from $e^1 + e^4 \leftrightarrow e^2 + e^3$, $e^1 - e^4$ and $e^2 - e^3$ unchanged we get M_{17}, M_{28}, M_{35} and M_{46}; and M_{27}, M_{18}, M_{45} and M_{36} which mix.

There is another way of looking at this automorphism, in terms of the regular maximal subgroup of SO(8), SU(2) x SU(2) x SU(2) x SU(2). It is instructive both because it gives some insight into the structure of the group and because it is a nice example of the use of extended Dynkin diagrams.

Recall that to form the extended Π-system, we add the lowest root α^0. For SO(8) the extended Π-system looks

like

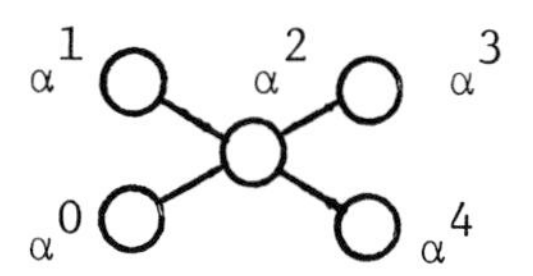

(XXViv)

where the lowest root is $\alpha^0 = -e^1 - e^2$. Clearly, the only way to obtain a nontrivial regular maximal subalgebra by removing a root is to remove α^2, leaving the 4 commuting SU(2) algebras associated with the orthogonal roots α^0, α^1, α^3 and α^4.

Consider the action of this subalgebra on the spinor representation D^3, whose weights are $1/2\ \eta_i e^i$ where

$$\prod_i \eta_i = -1.$$

These break up into sets corresponding to irreducible representations of the subalgebra. The set

$$\frac{1}{2}(e^1+e^2+e^3-e^4),\ \frac{1}{2}(e^1+e^2-e^3+e^4),$$

$$\frac{1}{2}(-e^1-e^2+e^3-e^4) \text{ and } \frac{1}{2}(-e^1-e^2-e^3+e^4) \tag{XXV.21}$$

transforms trivially under $E_{\pm\alpha^1}$ and $E_{\pm\alpha^4}$ while each weight is a component of a doublet under the SU(2)'s generated by $E_{\pm\alpha^0}$ and $E_{\pm\alpha^3}$. Similarly the set

$$\frac{1}{2}(e^1-e^2+e^3+e^4),\ \frac{1}{2}(e^1-e^2-e^3-e^4), \tag{XXV.22}$$

$$\frac{1}{2}(-e^1+e^2+e^3+e^4) \quad \text{and} \quad \frac{1}{2}(-e^1+e^2-e^3-e^4) \qquad \text{(XXV.22)}$$

transforms trivially under $E_{\pm\alpha^0}$ and $E_{\pm\alpha^3}$ and like doublets under $E_{\pm\alpha^1}$ and $E_{\pm\alpha^4}$. In a compact notation, we say that under the

$$SU(2)^0 \times SU(2)^1 \times SU(2)^3 \times SU(2)^4 \qquad \text{(XXV.23)}$$

subgroup, the spinor representation D^3 transforms like (2, 1, 2, 1) + (1, 2, 1, 2).

Obviously there are two other symmetrical possibilities for the breakdown of an 8:

$$(2, 2, 1, 1) + (1, 1, 2, 2)$$

and

$$(2, 1, 1, 2) + (1, 2, 2, 1). \qquad \text{(XXV.24)}$$

These correspond (in some order) to the vector representation D^1 and the other spinor representation D^4.

PROBLEMS FOR CHAPTER XXV

(XXV.A) Carry through the argument discussed at the end of the chapter and determine which representation in (XXV.25) is D^1 and which is D^4. Explain what the mappings (XXV.13) do to the four SU(2) factors.

(XXV.B) Does SO(8) have an SO(5) subgroup under which one spinor (D^4 say) transforms like two SO(5) spinors while the other spinor (D^3) transforms like an SO(5) vector and three singlets? Explain.

XXVI. Sp(2N)

Before discussing Sp(2N) in more detail, let us go back to the SU(N) subgroup and find a notation which is easier to work with. The weights of the N, ν^i for i=1 to N have the following properties (XIII.8):

$$(\nu^i)^2 = \frac{1}{2}\frac{N-1}{N}, \quad \nu^i \cdot \nu^i = -\frac{1}{2N} \quad i \neq j, \quad \sum_{i=1}^{N} \nu^i = 0. \tag{XXVI.1}$$

Howard Georgi, Lie Algebras in Particle Physics: From Isospin to Unified Theories ISBN 0-8053-3153-0

The last condition is obvious from the tracelessness of the SU(3) generators. The first two simply reflect the fact that the weights are symmetrical. The N ν^i's span an N-1 dimensional space, but we can make the above properties more obvious by embedding the ν^i's in an N dimensional space, as follows:

$$\nu^i = \frac{1}{\sqrt{2}}\,(e^i - \Sigma/N), \tag{XXVI.2}$$

where e^i are an orthonormal basis and

$$\Sigma = \sum_{i=1}^{N} e^i. \tag{XXVI.3}$$

In this notation, the simple roots of SU(N) are

$$\alpha^i = \nu^i - \nu^{i+1} = \frac{1}{\sqrt{2}}\,(e^i - e^{i+1}),\ i=1 \text{ to } N-1. \tag{XXVI.4}$$

These are also the first N-1 simple roots of Sp(2N). The last simple root is

$$\alpha^N = 2\nu^N + \sqrt{\frac{2}{N}}\,\nu^{N+1} \tag{XXVI.5}$$

where ν^{N+1} is a unit vector orthogonal to ν^i for $i = 1$ to N. Now we can take

$$\nu^{N+1} = \frac{1}{\sqrt{N}}\,\Sigma. \tag{XXVI.6}$$

Then,

$$\alpha^N = \sqrt{2}\; e^N . \tag{XXVI.7}$$

In fact, we could have written this down directly from the Dynkin diagram by comparison with SO(2N+1), but now we have made contact with our previous notation.

So the roots are

$$(\pm e^i \pm e^j)/\sqrt{2} \quad \text{for} \quad i \neq j \quad \text{and} \quad \pm\sqrt{2}\, e^i . \tag{XXVI.8}$$

The weights of the 2N dimensional representation are

$$\pm\, e^i/\sqrt{2} . \tag{XXVI.9}$$

Notice that the 2N dimensional representation is the fundamental representation with highest weight

$$\mu^1 = e^1/\sqrt{2} . \tag{XXVI.10}$$

The other fundamental representations have highest weight

$$\mu^j = \sum_{i=1}^{j} e^i/\sqrt{2} . \tag{XXVI.11}$$

Obviously, D^j (with highest weight μ^j) has something to do with the antisymmetric tensor product of j D^1's.

If we invent a tensor language in which the states of D^1 have a lower index, the tensor coefficients have upper indices and transform as follows

$$(T_a u)^i = T^i_{aj} u^j \tag{XXVI.12}$$

where T_a are the generators of the defining representation. They have the form

$$T^i_{aj} = (\vec{S} \cdot \vec{\sigma} + A)^i_j \tag{XXVI.13}$$

where the $\vec{S}$ are real symmetric NxN matrices, A is an antisymmetric imaginary NxN matrix and $\vec{\sigma}$ are the 2x2 Pauli matrices, as discussed in Chapter XIX.

The complex conjugate representation, with a lower index transforms as

$$(T_a v)_i = -T^j_{ai} v_j = -T^{i*}_{aj} v_j . \qquad \text{(XXVI.14)}$$

But this representation is pseudoreal, because

$$-\sigma_2 T^*_a \sigma_2 = T_a . \qquad \text{(XXVI.15)}$$

As we discussed in Chapter XXI (see (XXI.46-47)), this implies that the σ_2 is an invariant tensor with two upper indices, or two lower indices since it is its own inverse. Thus, given a tensor with a lower index, we can use σ_2 to raise it. Define

$$\tilde{v}^i = \sigma_2^{ij} v_j . \qquad \text{(XXVI.16)}$$

Note here that σ_2^{ij} and its inverse σ_{2ij} are 2Nx2N matrices, the product the Pauli matrix in the 2x2 space with the unit matrix in the NxN space. Then $\tilde{v}$ satisfies

$$(T_a \tilde{v})^i = \sigma_2^{ij} (T_a v)_j = -\sigma_2^{ij} T^k_{aj} v_k$$

$$= -\sigma_2^{ij} T^k_{aj} \sigma_{k\ell} \sigma^{\ell m} v_m = T^i_{a\ell} \tilde{v}^\ell . \qquad \text{(XXVI.17)}$$

Thus, we need only consider tensors with upper indices. We may sometimes want to include lower indices as a convenience, but they are equivalent to upper indices as in (XXVI.16).

Now with this tensor notation in hand, let us consider the antisymmetric tensor product $(D^1 \otimes D^1)_{AS}$ of two tensors u^i and v^j. It has components which are an antisymmetric tensor

$$u^i v^j - u^j v^i . \qquad \text{(XXVI.18)}$$

This is not quite an irreducible representation. We can use the invariant tensor σ_2 to reduce out the singlet,

$$\sigma_{2ij} u^i v^j . \tag{XXVI.19}$$

We can write (XXVI.18) as

$$w^{ij} + \frac{1}{N} \sigma_2^{ij} \sigma_{2k\ell} u^k v^\ell \tag{XXVI.20}$$

where

$$w^{ij} \equiv u^i v^j - u^j v^i - \frac{1}{N} \sigma_2^{ij} \sigma_{2k\ell} u^k v^\ell \tag{XXVI.21}$$

(Note that $\sigma_2^{ij} \sigma_{2ij} = 2N$). The combination w^{ij} transforms like the irreducible, fundamental representation D^2. You can prove this more rigorously (if you feel like it) by arguments analogous to (X.12-14).

We have argued that

$$(D^1 \otimes D^1)_{AS} = D^2 \oplus 1 . \tag{XXVI.22}$$

What about the symmetric combination, $(D^1 \otimes D^1)_S$? The symmetric product cannot be contracted with σ_2. It is irreducible. It has highest weight $2\mu^1$. You can check that the adjoint representation also has highest weight $2\mu^1 = \sqrt{2}\, e^1$, thus (XXVI.22) transforms like the adjoint representation. It does not look like the $(T_a)^i_j$, (XXVI.13), because these have one upper and one lower index. If we raise the lower index, the resulting matrices

$$T_a \sigma_2 = (\vec{S} \cdot \vec{\sigma} + A) \sigma_2 , \tag{XXVI.23}$$

do indeed comprise all the symmetric 2Nx2N matrices.

I will not go on with the tensor analysis of an arbitrary Sp(2N) representation. You can work it out for yourself with the twin tools of the fundamental weight idea and invariant tensor σ_2. I will stop here because I do not know of any important applications of Sp(2N) to particle physics.

PROBLEMS FOR CHAPTER XXVI

(XXVI.A) Find a set of Cartan generators for $\vec{\sigma}\cdot\vec{S}+A$ which makes (XXVI.10) obvious.

(XXVI.B) Find $D^1 \otimes D^2$ in Sp(6).

XXVII. LIE ALGEBRAS IN PARTICLE PHYSICS

I hope that the reader who has come this far feels comfortable with the simple Lie algebras as a tool for studying physics. We have studied some of the simplest and most generally useful examples. In this chapter, I will mention a few of the things we have not discussed in any detail and close with some comments on the future of Lie algebras in particle physics.

Howard Georgi, Lie Algebras in Particle Physics: From Isospin to Unified Theories ISBN 0-8053-3153-0

EXCEPTIONAL ALGEBRAS

The search for interesting unified theories, like the SU(5) and SO(10) theories, continues. One amusing theory is based on the algebra of E_6, one of the exceptional Lie algebras. The exceptional algebras are associated with the octonians, a peculiar set of objects of the form

$$a + b_\alpha i_\alpha \qquad \text{(XXVII.1)}$$

where a and b_α are real numbers and $\alpha = 1$ to 7. The i_α have the following multiplication law:

$$i_\alpha i_\beta = -\delta_{\alpha\beta} + g_{\alpha\beta\gamma} i_\gamma , \qquad \text{(XXVII.2)}$$

where $g_{\alpha\beta\gamma}$ is completely antisymmetric. In some basis, g is

$$g_{123} = g_{247} = g_{451} = g_{562} = g_{634} = g_{375} = g_{716} = 1, \qquad \text{(XXVII.3)}$$

with all other components either zero or obtainable from (XXVII.3) by antisymmetrization.

The algebra (XXVII.2) shares a nice property with the real numbers, the complex numbers and the quaternions (numbers of the form $a + i b_a \sigma_a$ where σ_a are Pauli matrices). The absolute value

$$|a + b_\alpha i_\alpha| = (a^2 + b_\alpha^2)^{1/2} \qquad \text{(XXVII.4)}$$

is preserved under multiplication. If A and B are octonians,

$$|AB| = |A||B|. \qquad \text{(XXVII.5)}$$

But the multiplication law is not associative. For example,

$$(i_1 i_2) i_7 = i_3 i_7 = i_5, \quad i_1(i_2 i_7) = i_1(-i_4) = -i_5. \qquad \text{(XXVII.6)}$$

I will not discuss the connection between these funny things and the exceptional algebras in any detail. But you can show, for example, that G_2 is the subgroup of SO(7) that treats $g_{\alpha\beta\gamma}$ as an invariant tensor.

But back to E_6. Note first that to go from $E_8 \to E_7$ or $E_7 \to E_6$, you remove one circle from the left branch of the Dynkin diagram in (XX.xvi). Continuing the same series, you can see that E_5 = SO(10) and E_4 = SU(5). Thus, since E_4 and E_5 give sensible unified theories, you might expect E_6 to be interesting. Certainly it contains SO(10) and SU(5) as regular subalgebras.

In fact, E_6 has a complex 27 dimensional representation which transforms under the SO(10) subgroup as a $D^5 \oplus D^1 \oplus 1$. Thus, it contains the representation that describes the creation operators for the right-handed particles. It also contains other stuff for which there is no experimental evidence.

The large exceptional algebras are even worse in the sense that they describe more unwanted (or at least, as yet unobserved particles). E_7 and E_8 have only real representations, hence they do not fit well with the light particles which transform according to a complex representation of SU(3) x SU(2) x U(1), (XVIII.8).

ANOMALIES

There is a peculiar constraint on unified theories that follows from the structure of quantum field theory, the mathematical language in which all these theories are based. The constraint is that if the creation operators for all the right-handed spin 1/2 particles transform according to a representation generated by matrices T_a^R, then T_a^R must satisfy

$$\mathrm{tr}(\{T_a^R, T_b^R\}T_c^R) = 0. \tag{XXVII.7}$$

You can show that this symmetric trace of three generators vanishes for all simple algebras except SU(N), N $\geq$ 3 (and SO(6) which is equivalent to SU(4)). In SU(N), suppose T_a^D generate the representation D of SU(N). Then define,

$$\mathrm{tr}(\{T_a^{D^1}, T_b^{D^1}\}T_c^{D^1}) \equiv d^{abc}, \tag{XXVII.8}$$

for the defining representation D^1. Then, for any representation, you can show that

$$\mathrm{tr}(\{T_a^D, T_b^D\}T_c^D) = A(D)d^{abc}, \tag{XXVII.9}$$

where A(D) is an integer called the <u>anomaly</u> of the representation D. Thus, (XXVII.7) is the statement that the creation operators for the right-handed particles transform according to an anomaly free representation of the unifying group.

You can derive the following properties of A(D):

$$A(\bar{D}) = -A(D); \tag{XXVII.10}$$

$$A(D_1 \oplus D_2) = A(D_1) + A(D_2); \tag{XXVII.11}$$

$$A(D_1 \otimes D_2) = A(D_1)\dim(D_2) + A(D_2)\dim(D_1). \tag{XXVII.12}$$

You can show by direct calculation that the representation (XVIII.8) of SU(3) x SU(2) x U(1) is anomaly free, and that it remains so when unified into SU(5).

THE FLAVOR PUZZLE

The interesting applications of Lie algebras to particle physics have changed their character in recent years. The approximate symmetries isospin, SU(3) and SU(6) evolved at a time when it seemed that more detailed dynamical information about the strong interactions was beyond our grasp. Now these symmetries, as well as others (the chiral symmetries which we have not discussed because they require a detailed knowledge of quantum field theory) are understood in terms of the underlying dynamics. The approximate symmetries are still useful and important, but they no longer seem as fundamental as they did in the sixties. It is amusing and almost ironic that the dynamics itself involves the same Lie algebras.

There are many puzzles left in particle physics in which Lie algebras may find applications. The deepest seems to be the flavor puzzle. Besides the u and d quarks and the electron and its neutrino which make up the SU(5) family (XVIII.15 and 18), there are heavier copies of the quark and the electron, and more light neutrinos. We have discussed the quarks, s, c and b. There is probably a t with charge 2/3. The heavy copies of the electron are called leptons. The muon and the τ have been seen so far.

While it is nice for particle physicists to have these heavy objects to study, we have absolutely no idea why Nature has chosen to repeat herself in this way. Is this evidence of some more profound unification?

PROBLEMS FOR CHAPTER XXVII

(XXVII.A) Construct simple roots and fundamental weights explicitly for E_6. Find a regular SU(5) subgroup of E_6 and determine the SU(5) representation to which the fundamental weights correspond.

(XXVII.B) Prove (XXVII.10-12).

(XXVII.C) Show that (XVIII.8) is anomaly free.

(XXVII.D) Calculate A(10) in SU(5). Hint: set a=b=c in (XXVII.8-9) and choose a convenient generator.

BIBLIOGRAPHY

BIBLIOGRAPHY

There are many excellent textbooks on general group theory to which the reader can refer when the idiosyncratic nature of this book gets to him or her. For example,

M. Hamermesh, Group Theory and its Application to Physical Problems (Addison-Wesley, Reading, Mass., 1962);

W. Miller, Symmetry Groups and their Applications (Academic Press, New York, 1972);

M. Tinkham, Group Theory and Quantum Mechanics (McGraw-Hill, New York, 1964).

On the more restricted subject of Lie algebras, the most reliable reference is the original paper by Dynkin,

E.B. Dynkin, American Mathematical Society Translations, Series 1, Vol. 9, pp. 328-469, and Series 2, Vol. 6, pp. 111-244.

There are also textbooks in which this material has been predigested for physicists,

R. Gilmore, Lie Groups, Lie Algebras and Some of their Applications (Wiley, New York, 1974).

B.G. Wybourne, Classical Groups for Physicists (Wiley, New York, 1974).

In the special case of SU(N), there are many excellent texts and reviews. Two of my favorites are

C. Itzykson and M. Nauenberg, Reviews of Modern Physics 38, 95 (1966);

D.B. Lichtenberg, Unitary Symmetry and Elementary Particles (Academic Press, New York, 1970).

More detailed discussion of hadron masses in a color SU(3) theory can be found in

A. De Rùjula, H. Georgi and S.L. Glashow, Physical Review D12, 147 (1975).

There is no good review of unified theories at exactly the level of the discussion in this book. For a nonmathematical review, see

H. Georgi, Scientific American, Vol. 244, 48 (April, 1981).

SUBJECT INDEX

SUBJECT INDEX